SEMINARS IN MATHEMATICS
V. A. STEKLOV MATHEMATICAL INSTITUTE, LENINGRAD

ZAPISKI NAUCHNYKH SEMINAROV
LENINGRADSKOGO OTDELENIYA
MATEMATICHESKOGO INSTITUTA IM. V. A. STEKLOVA AN SSSR

ЗАПИСКИ НАУЧНЫХ СЕМИНАРОВ
ЛЕНИНГРАДСКОГО ОТДЕЛЕНИЯ
МАТЕМАТИЧЕСКОГО ИНСТИТУТА им. В.А. СТЕКЛОВА АН СССР

SEMINARS IN MATHEMATICS
V. A. Steklov Mathematical Institute, Leningrad

1	Studies in Number Theory	A. V. Malyshev, Editor
2	Convex Polyhedra with Regular Faces	V. A. Zalgaller
3	Potential Theory and Function Theory for Irregular Regions	Yu. D. Burago and V. G. Maz'ya
4	Studies in Constructive Mathematics and Mathematical Logic, Part I	A. O. Slisenko, Editor
5	Boundary Value Problems of Mathematical Physics and Related Aspects of Function Theory, Part I	V. P. Il'in, Editor
6	Kinematic Spaces	R. I. Pimenov
7	Boundary Value Problems of Mathematical Physics and Related Aspects of Function Theory, Part II	O. A. Ladyzhenskaya, Editor
8	Studies in Constructive Mathematics and Mathematical Logic, Part II	A. O. Slisenko, Editor
9	Mathematical Problems in Wave Propagation Theory, Part I	V. M. Babich, Editor
10	Isoperimetric Inequalities in the Theory of Surfaces of Bounded External Curvature	Yu. D. Burago
11	Boundary Value Problems of Mathematical Physics and Related Aspects of Function Theory, Part III	O. A. Ladyzhenskaya, Editor
12	Investigations in the Theory of Stochastic Processes	V. N. Sudakov, Editor
13	Investigations in Classical Problems of Probability Theory and Mathematical Statistics, Part I	V. M. Kalinin and O. V. Shalaevskii
14	Boundary Value Problems of Mathematical Physics and Related Aspects of Function Theory, Part IV	O. A. Ladyzhenskaya, Editor
15	Mathematical Problems in Wave Propagation Theory, Part II	V. M. Babich, Editor
16	Studies in Constructive Mathematics and Mathematical Logic, Part III	A. O. Slisenko, Editor
17	Mathematical Problems in Wave Propagation Theory, Part III	V. M. Babich, Editor
18	Automatic Programming and Numerical Methods of Analysis	V. N. Faddeeva, Editor

SEMINARS IN MATHEMATICS
V. A. Steklov Mathematical Institute, Leningrad

Volume 15

MATHEMATICAL PROBLEMS IN WAVE PROPAGATION THEORY

Part II

Edited by V. M. Babich

Translated from Russian

c/b CONSULTANTS BUREAU • NEW YORK—LONDON • 1971

The original Russian text was published by Nauka Press in Leningrad in 1969 by offset reproduction of manuscript. The hand-written symbols have been retained in this edition. The present translation is published under an agreement with Mezhdunarodnaya Kniga, the Soviet book export agency.

В.М. БАБИЧ

МАТЕМАТИЧЕСКИЕ ВОПРОСЫ
ТЕОРИИ РАСПРОСТРАНЕНИЯ ВОЛН. 2

MATEMATICHESKIE VOPROSY TEORII
RASPROSTRANENIYA VOLN. 2

Library of Congress Catalog Card Number 79-13851
SBN 306-18815-5

A Division of Plenum Publishing Corporation
227 West 17th Street, New York, N. Y. 10011

United Kingdom edition published by Consultants Bureau, London
A Division of Plenum Publishing Company, Ltd.
Davis House (4th Floor), 8 Scrubs Lane, Harlesden, NW10 6SE, England

Printed in the United States of America

PREFACE

For some time now members of the Seminar on the Mathematical Theory of Wave Diffraction and Propagation at the Leningrad Section of the V. A. Steklov Mathematical Institute (LOMI) of the Academy of Sciences of the USSR have been working diligently to develop the parabolic equation method of Academicians V. A. Fok and M. A. Leontovich, as well as its extension commonly known as the method of standard problems. These methods have proved especially rewarding in application to the theory of open resonators. Several of the articles in the present collection are concerned with this topic.

In one article V. M. Babich investigates the multilayered resonator problem. A necessary and sufficient criterion is found for the stability of this type of resonator. An interesting result is the fact that the multilayered resonator is stable only if certain relations are met between its defining parameters; i.e., in general it is unstable. The parabolic equations for the eigenfunctions with appropriate boundary conditions can be solved if and only if the resonator is stable.

For the eigenfunctions of the Laplace operator the late M. F. Pyshkina,* a senior in the Physics Department of Leningrad State University (LGU), has studied their higher approximations concentrated in the neighborhood of a closed geodesic on an $(m+1)$-dimensional Riemann manifold. Following a notion expounded by V. F. Lazutkin, M. F. Pyshkina finds the wanted higher approximations for the eigenfunctions in the form of a series whose terms are linear combinations of the functions U_n (see the text of the article). The methods used by M. F. Pyshkina serve as the basis for the formulations of B. N. Semenov in a treatment of the higher approximations in the scalar problem of the three-dimensional multiple-mirror resonator, and those of T. F. Pankratova for solution of the same problem in the vector case. The eigenfrequencies exhibit differences in the vector and scalar cases already in the terms of order $o(1)$. This stems from the fact that in the problem studied by T. F. Pankratova it is impossible to distinguish the TE and TM modes.

Employing a procedure similar to that used by M. F. Pyshkina and B. N. Semenov, N. V. Svanidze finds a correction in explicit form to the first-approximation formula for the frequency of a three-dimensional double-mirror resonator. T. M. Popova, using the method of standard problems, analyzes the plane scalar case of the multiple-mirror resonator problem.

All of the foregoing papers dealing with the formulation of higher approximations make it possible to deduce mathematically rigorous results with regard to the eigenvalues and eigenfunctions of the problems in question. The abstract basis for this is found in the following theorem from the theory of operators in a Hilbert space:

Let A be a self-adjoint operator with discrete spectrum in a Hilbert space. Also, let $\lambda_1, \lambda_2, \ldots, \lambda_n \ldots \to \infty$ be a sequence of real numbers, and let $u_1, u_2, \ldots, u_n \ldots$ be a sequence of elements of the Hilbert space that are equal modulo one and are such that:

1) u_n belongs to the domain of definition of the operator A;

2) $\|(A-\lambda_n)u_n\| \leq C\lambda_n^{-s}$.

Then:

1) there exists a subsequence of eigenvalues of A:

$$\lambda^+_{x_1},\ \lambda^+_{x_2},\ \ldots$$

*M. F. Pyshkina lost her life tragically in February of 1969 while attempting to climb Mt. Belukha (in the Altai range) in midwinter. Her paper was prepared for publication by her academic sponsor, V. M. Babich.

such that

$$|\lambda_n - \lambda_{\varkappa_1}| < C\lambda_n^{-s};$$

2)

$$\|U_n - P_{\Delta_n} U_n\| \leqslant \frac{C}{C_1} \lambda_n^{-(s-s_1)},$$

where $P_{\Delta n}$ - is the projector onto the part of the spectrum in the interval

$$\Delta_n = [\lambda_n - C_1 \lambda_n^{-s}, \ \lambda_n + C_1 \lambda_n^{-s_1}].$$

It can be shown with the aid of this theorem that the formal series derived by M. F. Pyshkina, B. N. Semenov, T. F. Pankratova, and T. M. Popova for the eigenfrequencies are asymptotic expansions of the eigenvalues of certain boundary-value problems treated by the authors.

In an article by N. Ya. Kirpichnikova some very special asymptotic solutions are constructed for the dynamic equations of elasticity theory; at high frequencies they are concentrated in the neighborhood of an isolated ray of Rayleigh or transverse waves on the surface of the elastic body. The author's procedure is an interesting analog of the Olver method in the theory of linear ordinary differential equations containing a large parameter.

In an article by P. V. Krauklis the influence of the velocity characteristics of the media on the amplitudes of nonstationary Rayleigh and Stonely surface waves is assessed in the ray approximation. A method proposed earlier by V. M. Babich is used in the paper, whereby it is possible on the basis of energy relations to find the field characteristics left invariant along the rays of propagating elastic waves.

In another article V. M. Babich and Yu. P. Danilov construct a formal solution of the Schrödinger equation concentrated in the neighborhood of a certain classical path of motion.

Also bearing on the parabolic equation method is an article by N. V. Tsepelev. Here the author considers the problem of wave propagation in the shadow zone created at a first-order discontinuous interface between two inhomogeneous media. The velocities v_ν in the two media are such that with fixed, $v_1 = \text{const}$, $v_2(z)$ is a decreasing function, and $v_2(0) = v_1$.

An article by A. S. Blagoveshchenskii is devoted to the solution of the converse problem for a hyperbolic equation. The investigation is aimed at finding the coefficients of the equation from certain characteristics of the solution of the boundary-value problem studied in the article.

In an article by B. A. Belinskii a number of problems are studied in the determination of a pair of functions, one of which belongs to a certain subspace of the space $L_2(a, b)$, where $|a|$ and $|b| < \infty$, while the other is its orthogonal complement. A linear relation is assumed to link the two functions. Some of the problems treated are solved in explicit form, while others are reduced to equations of the Fredholm type.

V. M. Babich

CONTENTS

EIGENMODES OF A MULTILAYERED RESONATOR

V. M. Babich

In the present article we investigate the stability (see [1, 2]) of a system of rays in the neighborhood of the axis of a multilayered (see [3]) resonator. The fundamental result may be tersely stated as follows: The multilayered resonator, in general, is unstable. It will have stability only in the event that certain relations hold between the parameters characterizing it. Stability of the resonator is equivalent to the "compatibility condition" (considered by V. F. Boitsov [3]) and the possibility of finding the eigenfrequencies by the parabolic equation method (provided the parabolic equation method is applied in the manner customary for such problems). I am most grateful to R. G. Gordeev for helping me on the problems covered in the article. R. G. Gordeev provided me with the examples of unstable resonators. These examples then afforded the starting point for the present article.

§ 1. Statement of the Problem

Let the y axis be the axis of a multilayered resonator consisting of three layers.* The reflecting (refracting) boundaries are described up to high-order terms by the equations

$$y = \ell_i + \frac{1}{2\rho_i} x^2, \qquad i = 0, 1, 2, 3, \quad \ell_0 < \ell_1 < \ell_2 < \ell_3. \tag{1.1}$$

The significance of the quantities ℓ_i is clear from the diagram of Fig. 1, and $\frac{1}{\rho_i}$ are the curvatures of the reflecting (refracting) boundaries. We shall not impose any restrictions on the symbol $\frac{1}{\rho_i}$ $(i = 0, 1, 2, 3)$.

We shall assume that the wave propagation velocity in the -ith layer is equal to $c_i (= \text{const})$. We are interested in rays near the y axis, which are reflected and refracted at points of the curves (1.1) according to the laws of geometrical optics. We shall confine our treatment to the plane problem.

Let a ray

$$x = \alpha y + \beta \tag{2.1}$$

near the y axis (i.e., such that α and β are small) be reflected from the i-th boundary (1.1). We write the equation for the reflected ray in analogous form:

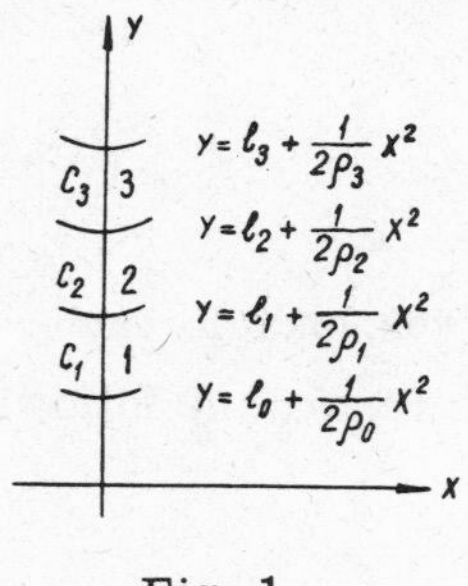

Fig. 1.

$$x = \hat{\alpha} y + \hat{\beta} \tag{3.1}$$

whereupon, up to high-order small terms (see [1, 2]),

$$\begin{pmatrix} \hat{\alpha} \\ \hat{\beta} \end{pmatrix} = \Gamma_j \begin{pmatrix} \alpha \\ \beta \end{pmatrix}, \tag{4.1}$$

*The method used here is applicable to a resonator consisting of any number of layers.

where Γ_j is a matrix having the form

$$\Gamma_j = \begin{pmatrix} -1-\dfrac{2\ell_i}{\rho_j}, & -\dfrac{2}{\rho_j} \\ 2\ell_j\left(1+\dfrac{\ell_j}{\rho_j}\right), & 1+\dfrac{2\ell_j}{\rho_j} \end{pmatrix}. \tag{5.1}$$

The following equation is geometrically obvious and easily verified analytically:

$$\Gamma_j = \Gamma_j^{-1}. \tag{6.1}$$

If the ray (2.1) is refracted at the boundary (1.1) and the equation for the reflected ray has the form (3.1), then up to principal terms

$$\begin{pmatrix} \hat{\alpha} \\ \hat{\beta} \end{pmatrix} = B_j \begin{pmatrix} \alpha \\ \beta \end{pmatrix}, \tag{7.1}$$

where

$$B_j = \begin{pmatrix} \dfrac{\ell_j}{\rho_j}\left(\dfrac{c_{j+1}}{c_j}-1\right)+\dfrac{c_{j+1}}{c_j} & \dfrac{1}{\rho_j}\left(\dfrac{c_{j+1}}{c_j}-1\right) \\ \ell_j\left(1-\dfrac{c_{j+1}}{c_j}\right)\left(1+\dfrac{\ell_j}{\rho_j}\right) & 1+\dfrac{\ell_j}{\rho_j}\left(1-\dfrac{c_{j+1}}{c_j}\right) \end{pmatrix}. \tag{8.1}$$

Let the ray $x = \alpha y + \beta$, traveling in the first medium from the boundary $y = \ell_1 + \frac{1}{2\rho_1}x^2$ to the boundary $y = \ell_0 + \frac{1}{2\rho_0}x^2$ be reflected and refracted several times, until finally it is transformed into the ray $x = \hat{\alpha}y + \hat{\beta}$. The path of the ray is illustrated schematically in Fig. 2. Up to principal terms we have

$$\begin{pmatrix} \hat{\alpha} \\ \hat{\beta} \end{pmatrix} = \Phi \begin{pmatrix} \alpha \\ \beta \end{pmatrix}, \tag{9.1}$$

where the matrix Φ has the form

$$\Phi = B_1^{-1}\Gamma_2\Gamma_1 B_2^{-1}\Gamma_3\Gamma_2\Gamma_3 B_2 (\Gamma_1\Gamma_2)^2 B_1 (\Gamma_0\Gamma_1)^2 \Gamma_0. \tag{10.1}$$

There are infinitely many rays analogous to the ray shown in Fig. 2, and there corresponds to each ray its own matrix Φ, which represents the product of matrices Γ_j and $B_j^{\pm 1}$. We regard the following as a natural definition.

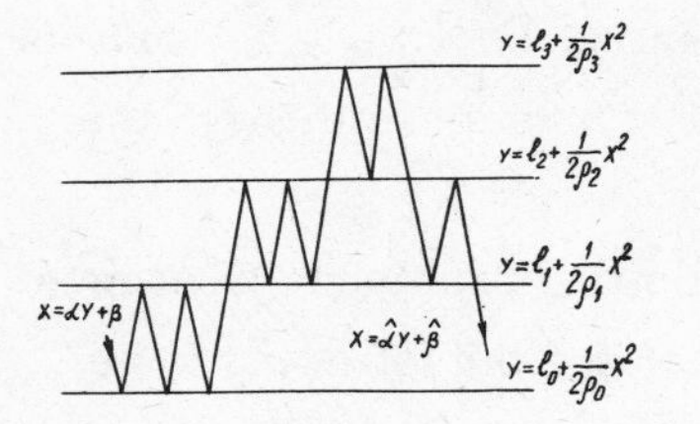

Fig. 2.

Definition. A multilayered resonator is said to be stable if all the matrices Φ corresponding to rays traveling in the resonator have norms* bounded in the aggregate, i.e., if there is a number $\mathcal{K}$ such that for any matrix, for the matrix (10.2) in particular,

$$\|\Phi\| \leqslant \mathcal{K}. \tag{11.1}$$

In this definition we confine ourselves to rays that originate in the first medium traveling "downward" and are transformed after all reflections and refractions once again into rays traveling "downward" in the first medium. We could investigate other analogous definitions of stability, for example, basing the definition on rays (and their corresponding matrices) originating in the second medium and ending in the third medium. It is readily seen, however, that all of these definitions lead to a condition equivalent to Eq. (11.1). We are concerned in this article with the derivation of necessary and sufficient conditions for stability of the resonator and with the explication of the relationship between stability of the resonator and the "compatibility condition" of V. F. Boitsov, as well as the solvability of the parabolic equation for the eigenmodes of the resonator.

§ 2. Algebraic Criterion of Resonator Stability

We denote by $\mathcal{M} = \{\Phi\}$ a set of matrices corresponding to rays proceeding "downward" in the first medium and becoming transformed as the ray of Fig. 2 after reflections and refractions once again into rays traveling "downward" in the first medium. We shall assume that the unit matrix is also included in $\mathcal{M}$. It is readily seen that the set of matrices $\mathcal{M}$ forms a monoid or semigroup (see [4], Chapt. I, § 1), i.e., a set:

1) in which multiplication is defined with the associativity property:

$$(\Phi'\Phi'')\Phi''' = \Phi'(\Phi''\Phi''');$$

2) which contains a "unit" element, i.e., an identity element e such that for all $\Phi \in \mathcal{M}$

$$e\Phi = \Phi e = \Phi.$$

Thus, it is apparent from geometrical considerations that the memberships $\Phi' \in \mathcal{M}$ and $\Phi'' \in \mathcal{M}$ imply $\Phi'\Phi'' \in \mathcal{M}$; the fulfillment of conditions 1) and 2) does not require any explanation.

The determinant of any matrix $\Phi \in \mathcal{M}$ is equal to one:

$$\det \Phi = 1 \tag{1.2}$$

(such matrices are often called unimodular matrices). This follows at once from:

1) the equation

$$\det \Gamma_j = -1, \tag{2.2}$$

which holds for any "reflective" matrix Γ_j (see § 1);

2) the fact that a ray initially traveling "downward" will once again travel "downward" if (and only if) it is reflected an even number of times, i.e., if the expression for Φ in terms of the matrices and B_j contains an even number of matrices Γ_j [see Fig. 2 and Eqs. (9.1) and (10.1)];

*Here we interpret the norm $\|\Phi\|$ of the matrix Φ in the usual sense as the minimum number of the set of numbers C such that

$$\left|\Phi\binom{\alpha}{\beta}\right| \leqslant C\sqrt{\alpha^2+\beta^2}$$

where α and β are any real numbers and $\left|\Phi\cdot\binom{\alpha}{\beta}\right|$ denotes the length of the vector $\Phi\binom{\alpha}{\beta}$ $\left(|\vec{\Psi}| = \left|\binom{\Psi_1}{\Psi_2}\right| = \sqrt{\Psi_1^2+\Psi_2^2}\right.$

3) the fact that the formulas for Φ, analogous to Eq. (10.1), contain the matrix B_j $(j = 1, 2)$ only as many times as the matrix B_j^{-1} (in order for the ray to return to medium 1 it must pass through the refracting boundary "upward" the same number of times as "downward").

Thus, the problem stated at the end of §1 is a special case of the following problem: What must be the monoid $\mathcal{M} = \{\Phi\}$ consisting of unimodular 2×2 matrices Φ in order for the following to be true for all $\Phi \in \mathcal{M}$:

$$\|\Phi\| \leq \mathcal{K} \tag{3.2}$$

($\mathcal{K}$ is independent of Φ)? We propose to solve this more general problem in the present section. If the set $\mathcal{M}$ were a group, then, as will become apparent shortly, it would be quite a simple matter to solve our problem, and we should direct our immediate effort toward reducing the problem to the analogous problem when the set of matrices satisfying condition (3.2) is a group.

We begin by proving Lemma 1.

Let $\mathcal{A}$ be a (2×2) real unimodular matrix, similar to an orthogonal matrix, i.e., stipulating the existence of a matrix C $(\det C \neq 0)$, such that

$$\mathcal{A} = C \begin{pmatrix} \cos\varphi & \sin\varphi \\ -\sin\varphi & \cos\varphi \end{pmatrix} C^{-1}, \tag{4.2}$$

$$\operatorname{Im} C = 0, \qquad 0 \leq \varphi < 2\pi;$$

then there exists a sequence of positive integers r_j $(j = 1, 2 \ldots,\ r_j \to +\infty)$ such that in the norm

$$\mathcal{A}^{r_j} \to \mathcal{A}^{-1}. \tag{5.2}$$

P r o o f. We examine two special cases:

1. Let $\frac{\varphi}{2\pi}$ be irrational. Then the set of fractional parts of the irrational numbers $\frac{r\varphi}{2\pi}$,... is dense on the interval (0,1) so that there is a subsequence $r_1, r_2, \ldots,$ such that

$$\left\{\frac{r_j\varphi}{2\pi}\right\} \to \frac{2\pi - \varphi}{2\pi} = \frac{\varphi_{-1}}{2\pi}; \quad \varphi_{-1} = 2\pi - \varphi \tag{6.2}$$

(where $\{G\}$ is the fractional part of G). Clearly,

$$\mathcal{A}^{r_j} = C \begin{pmatrix} \cos r_j\varphi, & \sin r_j\varphi \\ -\sin r_j\varphi, & \cos r_j\varphi \end{pmatrix} C^{-1} = C \begin{pmatrix} \cos\{r_j\varphi\}\cdot 2\pi, & \sin\{r_j\varphi\}\, 2\pi \\ -\sin\{r_j\varphi\}\cdot 2\pi, & \cos\{r_j\varphi\}\, 2\pi \end{pmatrix} C^{-1}$$

$$\xrightarrow[j\to\infty]{} C \begin{pmatrix} \cos\varphi_{-1} & \sin\varphi_{-1} \\ -\sin\varphi_{-1} & \cos\varphi_{-1} \end{pmatrix} C^{-1} = \mathcal{A}^{-1}$$

(note that for matrices of finite order convergence in the norm is equivalent to componentwise convergence).

2) Let

$$\frac{\varphi}{2\pi}=\frac{p}{q}; \quad p\geq 0, \quad q\geq 1,$$

where $\frac{p}{q}$ is an irreducible fraction. Then $\varepsilon = e^{i2\pi\frac{p}{q}}$ is the primitive root of q-th degree of 1 (see [5], Chapt. 4, §19). By virtue of the fact that $\varepsilon_{-1}=e^{i\varphi_{-1}}=e^{i2\pi\frac{q-p}{q}}$ is also a root of q-th degree of 1, for some integer $d\ (0\leq d\leq q-1)$ $\varepsilon^{d}=\varepsilon_{-1}$ or

$$\cos\varphi_{-1}=\cos\varphi d; \quad \sin\varphi_{-1}=\sin\varphi d.$$

Clearly, for the sequence τ_j we can adopt $\tau_j = d+jq \quad (j=0,1,2,\ldots)$. In this case

$$A^{\tau_j}=A^{-1}.$$

We now turn to the proof of a theorem that solves the problem stated at the outset of the section. In order for the 2×2 real unimodular matrices $\{\Phi\}$ forming the monoid $\mathcal{M}$ to have the property (3.2) it is necessary and sufficient that there exist a real matrix C, $\det C\neq 0$, that is independent of Φ and is such that

$$\Phi=CU_{\Phi}C^{-1}; \quad U_{\Phi}=\begin{pmatrix}\cos\varphi & \sin\varphi\\ -\sin\varphi & \cos\varphi\end{pmatrix}; \tag{7.2}$$

(φ depends on Φ).

Sufficiency is obvious in that

$$\|\Phi\|=\|CU_{\Phi}C^{-1}\|\leq\|C\|\,\|U_{\Phi}\|\,\|C^{-1}\|=\|C\|\cdot\|C^{-1}\|.$$

For the proof of necessity we let the monoid $\mathcal{M}=\{\Phi\}$ have the properties indicated in the statement of the theorem. Then any matrix $\Phi\in\mathcal{M}$ is similar to an orthogonal matrix

$$\Phi=C_{\Phi}U_{\Phi}C_{\Phi}^{-1}; \quad U_{\Phi}=\begin{pmatrix}\cos\varphi & \sin\varphi\\ -\sin\varphi & \cos\varphi\end{pmatrix};$$

otherwise, as we readily perceive, the sequence of powers Φ^{m}, $m=1,2,3\ldots$ would not have uniformly bounded norms, whereas, clearly, $\Phi^{m}\in\mathcal{M}$. We now form the group $\mathcal{N}=\{\Psi\}$, generated by matrices of the monoid $\mathcal{M}$, i.e., the set of all possible products of the form

$$\Psi=\Phi_1^{q_1}\Phi_2^{q_2}\cdots\Phi_m^{q_m}; \quad \Phi_j\in\mathcal{M}, \tag{8.2}$$

where q_j are integers, not necessarily positive. It may be assumed without loss of generality that $q_j=\pm 1$. It is seen at once that the matrices Ψ will satisfy inequality (3.2). Thus, if all $q_j=1$, then $\Psi\in\mathcal{M}$, and (3.2) holds. Now let certain $q_j=-1$. Let us assume, for example, that $\Psi=\Phi_1\Phi_2^{-1}\Phi_3$. Let τ_j be a sequence of positive integers such that

$$\Phi_2^{\tau_j}\longrightarrow\Phi_2^{-1}$$

(see the lemma).

Then

$$\|\Psi\| = \|\Phi_1 \Phi_2^{-1} \Phi_3\| = \lim_{j\to\infty} \|\Phi_1 \Phi_2^{\nu_j} \Phi_3\| \leq \mathcal{K}.$$

The same reasoning plainly carries over to the general case.

We denote by $\bar{\mathcal{N}}$ the closure of the group $\mathcal{N}$, i.e., the limits in the norm of all subsequences of matrices $\Psi \in \mathcal{N}$. The set of matrices $\bar{\mathcal{N}}$ will also be a group with the group multiplication operation, and obviously for the matrices forming $\bar{\mathcal{N}}$ inequality (3.2) will also hold.

The group $\bar{\mathcal{N}}$ is a compact group of real matrices. All linear representations of this group are equivalent to some unitary representation.* In particular, there is a real matrix C, identical for all matrices $\Psi \in \bar{\mathcal{N}}$, such that

$$\Psi = C U_\psi C^{-1}; \qquad U_\psi = \begin{pmatrix} \cos\varphi & \sin\varphi \\ -\sin\varphi & \cos\varphi \end{pmatrix}.$$

Considering that the monoid $\mathcal{M} \subset \bar{\mathcal{N}}$, we have thus proved our theorem.

Note that if the matrix C is replaced by $C\begin{pmatrix} 0 & 1 \\ 1 & 0 \end{pmatrix}$, the latter will also reduce all matrices $\Psi \in \bar{\mathcal{N}}$ to orthogonal form (merely replacing all φ by $-\varphi$). From the theorem just proved we can deduce effective stability criteria for the multilayered resonator.

Corresponding to rays that do not escape the first medium are matrices of the form $(\Gamma_1 \Gamma_0)^s$. By the above theorem we have

$$\Gamma_1 \Gamma_0 = C U_{10} C^{-1} \tag{9.2}$$

(where $\operatorname{Im} C = 0$; $\operatorname{Im} U_{10} = 0$, and U_{10} is orthogonal). We know, of course, that this condition is sufficient for the stability of a single-layer resonator. Any matrix corresponding to rays that do not escape the second medium is the product of matrices corresponding to rays "visiting" medium I only twice, once at the beginning and once at the end of the path. Corresponding to these rays are matrices of the form

$$(\Gamma_1 \Gamma_0)^{s_3} B_1^{-1} \Gamma_2 (\Gamma_1 \Gamma_2)^{s_2} B_1 (\Gamma_0 \Gamma_1)^{s_1} \Gamma_0 . \tag{10.2}$$

Assuming here that $s_1 = s_2 = s_3 = 0$, we find that

$$\Gamma_2^{B_1} \Gamma_0 = C U_{20} C^{-1} \tag{11.2}$$

(U_{20} is an orthogonal matrix, and $\Gamma_2^{B_1} = B_1^{-1} \Gamma_2 B_1$). Putting $s_1 = s_3 = 0$ and $s_2 = 1$ in Eq. (10.2), we find that

$$\Gamma_2^{B_1} \Gamma_1^{B_1} \Gamma_2^{B_1} \Gamma_0 = C U C^{-1}$$

(U is an orthogonal matrix, and $\Gamma_1^{B_1} = B_1^{-1} \Gamma_1 B_1$). From this, using (11.2) we obtain

$$\Gamma_2^{B_1} \Gamma_1^{B_1} = C U_{21} C^{-1}, \tag{12.2}$$

where U_{21} is an orthogonal matrix.

By virtue of the fact that any matrix of the form (10.2) can be represented as the product of matrices

---

*We are relying here on the fundamental postulates of the theory of linear representations of groups; see, e.g., [6].

$$\Gamma_1\Gamma_0, \quad \Gamma_2^{B_1}\Gamma_0, \quad \Gamma_2^{B_1}\Gamma_1^{B_1}, \tag{13.2}$$

we easily come to the conclusion that conditions (9.2), (11.2), and (12.2) are necessary and sufficient for the stability of a double-layer resonator. It is not too difficult to show that of the three necessary and sufficient stability conditions (9.2), (11.2), and (12.2), only two need remain, either (9.2) and (11.2) or (9.2) and (12.2). This follows at once from the identities

$$\Gamma_1^{B_1}\Gamma_1 = \begin{pmatrix} 1 & 0 \\ 0 & 1 \end{pmatrix}$$

and

$$(\Gamma_2^{B_1}\Gamma_1^{B_1})(\Gamma_2^{B_1}\Gamma_0)(\Gamma_1\Gamma_0)^{-1} = \Gamma_1^{B_1}\Gamma_1 = \begin{pmatrix} 1 & 0 \\ 0 & 1 \end{pmatrix}.$$

Necessary and sufficient conditions for the stability of a three-layer resonator are derived analogously. For the stability of a three-layer resonator it is necessary and sufficient that a double-layer resonator be stable [in other words, that Eqs. (9.2) and (11.2) or (9.2) and (12.2) be fulfilled] and, in addition, that either of the following two equations hold:

$$\Gamma_3^{B_1 B_2}\Gamma_1^{B_1} = C\,U_{31}\,C^{-1} \tag{14.2}$$

$$\Gamma_3^{B_1 B_2}\Gamma_2^{B_1 B_2} = C\,U_{32}\,C^{-1} \tag{15.2}$$

($\Gamma_3^{B_1 B_2} = B_1^{-1} B_2^{-1} \Gamma_3 B_2 B_1$ are orthogonal matrices). The stability criteria expressed by Eqs. (9.2)-(15.2) are still not adequately effective, but from them we can readily derive criteria at once amenable to verification. For the derivation of these criteria the following lemma is useful: Let the matrix* C reduce the unimodular matrix A to orthogonal form (i.e., $A = C U_A C^{-1}$, where U_A is an orthogonal matrix). In order for the unimodular matrix B to be reduced to orthogonal form by the same matrix C (i.e., in order to have $B = C U_B C^{-1}$, where U_B is orthogonal) it is necessary and, for $A \neq \begin{pmatrix} \pm 1 & 0 \\ 0 & \pm 1 \end{pmatrix}$, sufficient that A and B commute.

The necessity clause follows from the commutivity of orthogonal 2×2 matrices. The sufficiency clause follows from the readily verified fact that any unimodular matrix that commutes with an orthogonal matrix $\neq \pm \begin{pmatrix} 1 & 0 \\ 0 & 1 \end{pmatrix}$ is itself orthogonal.

Consequently, for a stable double-layer resonator the matrices (9.2) and (11.2) [or (9.2) and (12.2)] commute. Inasmuch as the coefficients of these matrices are expressed in terms of the parameters ρ_j, ℓ_j, and c_j describing the resonator, there must be certain relations between those parameters. It would appear that commutivity of the matrices (9.2) and (11.2) leads to four formulas relating ρ_0, ρ_1, ρ_2, ℓ_0, ℓ_1, ℓ_2, c_0, and c_1, but these four formulas reduce to a single one. The stability of the three-layer resonator yields still another equation involving, in addition to the parameters indicated above, c_2, ρ_2, and ℓ_2 as well.

We shall not write out these relations, as they can be obtained in a simpler form by alternative arguments.

§ 3. Closed Congruences of Rays in a Stable Multilayered Resonator

From now on we shall consider only stable resonators. Let $\varkappa > 0$ be any positive number. Consider a set of rays in the first medium, whose parameters α and β are connected by the relation

$$(C^{-1}\vec{\delta}_1, \; C^{-1}\vec{\delta}_1) = \varkappa. \tag{1.3}$$

*All of the matrices considered here are real 2×2 matrices.

Here C is a matrix reducing the matrices (9.2)-(15.2) to orthogonal form, and $\vec{\delta}_1 = \binom{\alpha}{\beta}$ is a vector with components α and β. For rays refracted from the first medium into the second (or from the second into the third medium) the corresponding vector $\vec{\delta}_2$ (or $\vec{\delta}_3$) is related to $\vec{\delta}_1$ by the formula

$$\vec{\delta}_2 = B_1 \vec{\delta}_1$$
$$(\text{or } \vec{\delta}_3 = B_2 B_1 \vec{\delta}_1) \tag{2.3}$$

If rays complying with relation (1.3) are refracted into the second and then into the third medium, the following formulas will hold in the given media for the vectors $\vec{\delta}_2$ and $\vec{\delta}_3$ defining these rays:

$$(C^{-1}B_1^{-1}\vec{\delta}_2,\ C^{-1}B_1^{-1}\vec{\delta}_2) = \varkappa, \tag{3.3}$$

$$(C^{-1}B_1^{-1}B_2^{-1}\vec{\delta}_3,\ C^{-1}B_1^{-1}B_2^{-1}\vec{\delta}_3) = \varkappa. \tag{4.3}$$

If a particular ray satisfying relation (1.3) returns to medium 1 after a series of refractions and reflections, it will be characterized by the vector $\vec{\delta}' = \Phi\vec{\delta}_1$, where in light of what was proved in the preceding section

$$\Phi = C U_\Phi C^{-1}, \tag{5.3}$$

where U_Φ is an orthogonal matrix. From (5.3) we infer

$$(C^{-1}\vec{\delta}',\ C^{-1}\vec{\delta}') = (C^{-1}\Phi\vec{\delta}_1,\ C^{-1}\Phi\vec{\delta}_1) = (U_\Phi C^{-1}\vec{\delta}_1,\ U_\Phi C^{-1}\vec{\delta}_1) = (C^{-1}\vec{\delta}_1,\ C^{-1}\vec{\delta}_1) = \varkappa,$$

i.e., the ray, on returning, is again converted to a ray belonging to the same set (1.3). A ray of the set (1.3), on being refracted from the first into the second medium, goes over to the congruence of rays (3.3). It is also invariant; no matter how many times a ray of the congruence (3.3) is reflected and refracted, on returning to medium 2 it will satisfy Eq. (3.3). An analogous invariant property is inherent in the congruence (4.3), which we shall not discuss further. Thus, any ray of the congruence (1.3), on being refracted and reflected, will still belong to the set of rays characterized by Eqs. (1.3), (3.3), and (4.3). Analogous sets of rays in problems involving open resonators are sometimes called closed congruences of rays. A closed congruence of rays can be specified by means of the Floquet solutions, whereupon it is possible to use a procedure originally "geared" to modeling problems [7, 8] and applied subsequently with success to the theory of open resonators [9, 10]. Let a ray $x = \alpha y + \beta$ be reflected from the zeroth, and then from the first boundary, changing into the ray $x = \alpha' y + \beta'$.

From the foregoing considerations we obtain the equations

$$(C^{-1}\vec{\delta},\ C^{-1}\vec{\delta}) = (C^{-1}\vec{\delta}',\ C^{-1}\vec{\delta}') \tag{6.3}$$
$$\vec{\delta}' = \Gamma_1\Gamma_0\vec{\delta}.$$

Vectors of identical length are derivable one from another by rotation, i.e., there is an orthogonal matrix

$$U_1 = \begin{pmatrix} \cos\varphi_1 & \sin\varphi_1 \\ -\sin\varphi_1 & \cos\varphi_1 \end{pmatrix}, \quad 0 \leq \varphi_1 < 2\pi, \tag{7.3}$$

such that

$$C^{-1}\vec{\delta}' = U_1 C^{-1}\vec{\delta}.$$

This formula, coupled with the arbitrariness of the vector $\vec{\delta}$ and Eq. (6.3), leads to the relation

$$C^{-1}\Gamma_1\Gamma_0 = U_1 C^{-1},$$

which is reduced by a simple identity transformation to the form

$$\Gamma_1\Gamma_0(\vec{C}_1 + i\vec{C}_2) = e^{i\varphi_1}(\vec{C}_1 + i\vec{C}_2),$$

where $\vec{C}_1$ and $\vec{C}_2$ are vectors whose components form the columns of the matrix C, i.e., the columns of C are the real and imaginary parts of the eigenvector of $\Gamma_1\Gamma_0$ corresponding to the eigenvalue $e^{i\varphi_1}$, where φ_1 is the number defining the matrix U_{φ_1} [see (7.3)]. The same matrix C reduces the matrix (11.2) to orthogonal form:

$$\Gamma_2^{B_1}\Gamma_1^{B_1} = B_1^{-1}\Gamma_2\Gamma_1 B_1, \tag{8.3}$$

so that $\vec{C}_1 + i\vec{C}_2 = \vec{\xi}_1$ is also an eigenvector of the matrix (8.3) corresponding to some eigenvalue $e^{i\varphi_2}$, $0 \leq \varphi_2 < 2\pi$:

$$B_1^{-1}\Gamma_2\Gamma_1 B_1(\vec{C}_1 + i\vec{C}_2) = e^{i\varphi_2}(\vec{C}_1 + i\vec{C}_2),$$

i.e., the vector $B_1(\vec{C}_1 + i\vec{C}_2) = \vec{\xi}_2$ is an eigenvector of the matrix $\Gamma_2\Gamma_1$. Consequently, if the complex "ray" $x = \alpha y + \beta$ (which does not have immediate physical meaning),

$$\begin{pmatrix}\alpha\\ \beta\end{pmatrix} = \vec{C}_1 + i\vec{C}_2 = \vec{\xi}_1 \tag{9.3}$$

is formally refracted from the first medium into the second, we obtain a ray characterized by an eigenvector of the matrix $\Gamma_2\Gamma_1$. Refracting the complex "ray" characterized by the vector

$$B_1(\vec{C}_1 + i\vec{C}_2) = \xi_2 \tag{10.3}$$

from the second into the third medium, we obtain a "ray" characterized by the vector

$$B_2 B_1(\vec{C}_1 + i\vec{C}_2) = \vec{\xi}_3, \tag{11.3}$$

which is an eigenvector of the matrix $\Gamma_3\Gamma_2$. Let us now consider in the j-th medium, $j = 1, 2, 3$, the complex "ray" $x = \alpha_j y + \beta_j = \gamma_j(y)$, where α_j and β_j are the components of the eigenvector ξ of the matrix $\Gamma_j\Gamma_{j-1}$ [see Eqs. (9.3)-(11.3)]. If the "ray" $x = \gamma_j(y)$ is reflected from the $(j-1)$st mirror and then from the j-th, it transforms into the ray $x = e^{i\varphi_j}\gamma_j(y)$, i.e., the "ray" $x = \gamma_j(y)$ plays the role of the Floquet solution for the j-th medium (cf. [7-9]), while the role of the closed path is taken by the twice-covered segment $\ell_{j-1} \leq y \leq \ell_j$ or, more precisely, two copies of the segment $\ell_{j-1} \leq y \leq \ell_j$ glued at the points $x = \ell_{j-1}$ and $y = \ell_j$. The functions $\gamma_j(y)$ can be used to write the closed congruence of rays in our problem in the form of three equations:

$$x = \sqrt{\frac{\varkappa}{2}}\,\mathrm{Re}\left(e^{i\mu_j}\gamma_j(y)\right) \tag{12.3}$$

$$j = 1, 2, 3 \qquad 0 \leq \mu_j < 2\pi.$$

The three equations (12.3) are completely equivalent to Eqs. (1.3), (3.3), and (4.3).

Moreover, the presence of Eqs. (12.3) is equivalent to stability of a resonator. Thus: Stability in a small multilayered resonator has brought us to Eqs. (12.3). Now let Eqs. (12.3) be valid, i.e., let every "partial" resonator

$$l_{j-1}+\frac{1}{2\rho_{j-1}}x^2 \le y \le l_j+\frac{1}{2\rho_j}x^2 \qquad (13.3)$$

be stable, resonator (13.3) corresponding to Floquet solutions $\gamma_j(y)$, such that by refracting the "ray" $x=\alpha_j y+\beta_j=\gamma_j(y)$ from the j-th into the $(j+1)$st medium, i.e., going from the ray $x=\alpha_j y+\beta_j$ to the ray $x=\alpha_{j+1}y+\beta_{j+1}$ where

$$\begin{pmatrix}\alpha_{j+1}\\ \beta_{j+1}\end{pmatrix}=B_j\begin{pmatrix}\alpha_j\\ \beta_j\end{pmatrix} \quad (B_j \text{ - refracting matrix})$$

we obtain the Floquet solution $\gamma_j(y)=\alpha_{j+1}y+\beta_{j+1}$ for the $(j+1)$st medium. It is also assumed that the real and imaginary parts of the vector $\begin{pmatrix}\alpha_j\\ \beta_j\end{pmatrix}=C_{1j}+iC_{2j}$ are linearly independent.

Now let the ray $x=\alpha y+\beta$ travel "downward" in the first medium to the zeroth boundary. By the proper choice of $\varkappa$ and μ_1 we can represent this vector in the form (12.3) (for $j=1$), a fact that is readily deduced from the linear independence of the real and imaginary parts of the vector $\vec{\xi}_1=C_1+iC_2$. A calculation shows that

$$\sqrt{\frac{\varkappa}{2}}=\|C^{-1}\vec{\delta}\|; \quad \vec{\delta}=\begin{pmatrix}\alpha\\ \beta\end{pmatrix}. \qquad (14.3)$$

After an arbitrary number of refractions and reflections the ray, on returning to medium 1, will again be described by Eq. (12.3) with exactly the same $\varkappa$ [see Eq. (13.3)] and, in general, a different μ_1 $(0\le\mu_1<2\pi)$. Thus, a ray characterized by the vector $\vec{\delta}$ goes over to a ray characterized by the vector

$$\begin{pmatrix}\alpha'\\ \beta'\end{pmatrix}=\|C^{-1}\vec{\delta}\|(\vec{C}_1\cos\mu-\vec{C}_2\sin\mu)=\|C^{-1}\vec{\delta}\|\cdot C\cdot\begin{pmatrix}\cos\mu\\ -\sin\mu\end{pmatrix},$$

$$\vec{C}_1+\vec{C}_2 i=\vec{\xi}_1;$$

C is a 2×2 matrix with columns $\vec{C}_1$ and $\vec{C}_2$. The length of the vector $\begin{pmatrix}\alpha'\\ \beta'\end{pmatrix}$ is clearly estimated by the quantity

$$\|C^{-1}\|\cdot\|C\|\cdot|\vec{\delta}|,$$

whence it follows that all the matrices of the monoid $\mathcal{M}=\{\Phi\}$ (see §§1, 2) have norms no greater than $\|C\|\cdot\|C^{-1}\|$. In this way we have inferred the stability of the system of rays from Eqs. (12.3).

§4. Solution of the Parabolic Equation

In the j-th medium let the Helmholtz equation be satisfied:

$$\Delta u+k_j^2 u=0 \qquad k_j=\frac{\omega}{c_j}. \qquad (1.4)$$

At an interface the usual continuity conditions are met on the function u and its derivative, whereas at reflecting boundaries the function u vanishes.

We propose to seek the function u in the j-th medium in the form of a superposition of two waves: the wave u_j^+ (or u_j^-) propagating "upward" (or "downward"), where, as usual, we assume that

$$u_j^{\pm} = e^{\pm i\kappa_j y} v_j^{\pm}; \qquad \kappa_j = \frac{\omega}{c_j}, \tag{2.4}$$

where $\nu_j^{\pm}$ are the attenuation functions (in the terminology of V. A. Fok), which vary more slowly than the oscillating factor $\exp(\pm i\kappa_j y)$.

Substituting expressions (2.4) into Eq. (1.4) and rejecting $\frac{\partial^2 \nu_j^{\pm}}{\partial y^2}$ (as is customarily done in problems of this nature), we arrive at the parabolic equations

$$\begin{gathered} \pm 2i\kappa_j \frac{\partial \nu_j^{\pm}}{\partial y} \pm \frac{\partial^2 \nu_j^{\pm}}{\partial x^2} = 0 \\ \pm 2i\kappa_j \frac{\partial \nu_j^{\pm}}{\partial y} + \frac{\partial^2 \nu_j^{\pm}}{\partial x^2} = 0. \end{gathered} \tag{3.4}$$

It is well known [8, 10] that Eqs. (3.4) are satisfied by the functions

$$\nu_j^{\pm} = \frac{\text{const}}{\sqrt{\gamma}} e^{\pm \frac{i}{2} \frac{\gamma'}{\gamma} \kappa_j x^2}, \tag{4.4}$$

where $\gamma = \alpha y + \beta$ is an arbitrary linear function of y. For the function $\gamma(y)$ for the j-th medium we pick the "Floquet solution" $\gamma_j(y)$ (see §3). We recall that by "refracting" the ray $x = \gamma_j(y)$ from the j-th medium into the $(j+1)$st $(j = 1, 2)$ we obtain the ray $x = \gamma_{j+1}(y)$ (see § 3). We are concerned with solutions that tend to zero with distance from the ray $x = 0$. This requires that

$$\operatorname{Im} \pm \frac{\gamma_j^{\pm\prime}}{\gamma_j^{\pm}} = \mp \frac{\operatorname{Im} \gamma_j \frac{d}{dy} \operatorname{Re} \gamma - \operatorname{Re} \gamma \frac{d}{dy} \operatorname{Im} \gamma}{|\gamma^{\pm}|^2} > 0$$

on the ray (the plus sign on the left-hand side corresponds to a wave traveling "upward," the minus sign to a wave traveling "downward"). In the case of a wave traveling "downward" the numerator on the right-hand side coincides with the determinant of the matrix C for the first medium (or with the determinants of the matrices $B_1 C$ or $B_2 B_1 C$ for the second or third medium, respectively). We are faced with the requirement of choosing C so that $\det C$ will be positive. Since C can always be replaced by $C \begin{pmatrix} 0 & 1 \\ 1 & 0 \end{pmatrix}$, this choice is always possible. If the above inequality holds for the incident wave, it also holds for the reflected wave, as is readily verified on the grounds that the determinant of the reflection matrix is equal to −1.

We begin with the first medium. Let a wave traveling "downward" (or "upward") be described by the expression

$$C_0^{-} \frac{1}{\sqrt{\gamma_1^{-}}} e^{-\frac{i}{2} \frac{\gamma_1^{-\prime}}{\gamma_1^{-}} \kappa_1 x^2}, \tag{5.4}$$

or

$$C_0^{+} \frac{1}{\sqrt{\gamma_1^{+}}} e^{\kappa_1 x^2 + i\kappa_1 y}, \tag{6.4}$$

where

$$\gamma_1^{\pm} = \alpha_1^{\pm} y + \beta_1^{\pm}.$$

Inasmuch as γ_1 is a Floquet solution (see § 3), the vector $\begin{pmatrix}\alpha_1^-\\ \beta_1^-\end{pmatrix}$ is an eigenvector of the matrix $\Gamma_1\Gamma_0$, whereas

$$\begin{pmatrix}\alpha_1^+\\ \beta_1^+\end{pmatrix} = \Gamma_0 \begin{pmatrix}\alpha_1^-\\ \beta_1^-\end{pmatrix} \tag{7.4}$$

(where Γ_0 is the reflection matrix), at the reflecting boundary $y = \ell_0 + \frac{1}{2\rho_0}x^2$ the following conditions hold:

$$\gamma_1^- = \gamma_1^+; \quad \frac{\gamma_1^{-1}}{\gamma_1^-} + \frac{1}{\rho_0} = -\frac{\gamma_1^{+1}}{\gamma_1^+} - \frac{1}{\rho_0}. \tag{8.4}$$

(The first equation states the requirement that the incident and reflected rays pass through the same point of the boundary $y = \ell_0 + \frac{1}{2\rho_0}x^2$, and the second expresses the coincidence of the angles of incidence and reflection.) Writing relations (8.4) in the form of a linear relation between the vectors (α_1^-, β_1^-) and (α_1^+, β_1^+), we arrive at once at expression (5.1) ($j=0$ for Γ_0). The addition of the incident (U_1^-) and the reflected (U_1^+) wave at $y = \ell_0 + \frac{1}{2\rho_0}x^2$ must yield a sum of zero, i.e.,

$$\frac{C_0^-}{\sqrt{\gamma_1^-}} e^{-\frac{i}{2}\frac{\gamma_1^{-\prime}}{\gamma_1^-}\kappa_1 x^2 - i\kappa_1 y} + \frac{C_0^+}{\sqrt{\gamma_1^+}} e^{\frac{i}{2}\frac{\gamma_1^{+\prime}}{\gamma_1^+}\kappa x^2 + i\kappa_1 y} = 0. \tag{9.4}$$

By virtue of relations (8.4) and the equation

$$y = \ell_0 + \frac{1}{2\rho_0}x^2$$

Eq. (9.4) is equivalent to a linear relation between C_0^+ and C_0^-:

$$C_0^- e^{-i\kappa_1\ell_0} + C_0^+ e^{i\kappa_1\ell_0} = 0. \tag{10.4}$$

Next we consider the refracting boundary

$$y = \ell_1 + \frac{1}{2\rho_1}x^2. \tag{11.4}$$

The ray $x = \alpha_1^+ y + \beta_1^+$, where

$$\begin{pmatrix}\alpha_1^+\\ \beta_1^+\end{pmatrix} = \Gamma_0 \begin{pmatrix}\alpha_1^-\\ \beta_1^-\end{pmatrix} \tag{12.4}$$

[$\begin{pmatrix}\alpha_1^-\\ \beta_1^-\end{pmatrix}$ is an eigenvector of the matrix $\Gamma_1\Gamma_0$], on being reflected from the boundary (11.4), is converted into a ray characterized by the vector

$$\Gamma_1 \Gamma_0 \begin{pmatrix} \alpha_1^- \\ \beta_1^- \end{pmatrix}. \tag{13.4}$$

Refracting the rays characterized by the vectors (12.4) and (13.4) into medium 2, we obtain rays characterized by the respective vectors

$$B_1 \Gamma_0 \begin{pmatrix} \alpha_1^- \\ \beta_1^- \end{pmatrix}, \quad B_1 \Gamma_1 \Gamma_0 \begin{pmatrix} \alpha_1^- \\ \beta_1^- \end{pmatrix}.$$

The ray corresponding to the vector

$$B_1 \Gamma_1 \Gamma_0 \begin{pmatrix} \alpha_1^- \\ \beta_1^- \end{pmatrix}$$

proceeds to the boundary, where it is reflected and transformed into the ray corresponding to the vector

$$B_1 \Gamma_0 \begin{pmatrix} \alpha_1^- \\ \beta_1^- \end{pmatrix}$$

(recall that the matrices Γ_1 and B_1 commute).

Now let the solutions corresponding to our rays by Eq. (4.4) (each $\gamma = \alpha y + \beta$ corresponds one-to-one to a ray $x = \gamma = \alpha y + \beta$) satisfy the following boundary conditions: continuity of the wave field and continuity of its normal derivative.

At the boundary (11.4) we have

$$\gamma_1^+ = \gamma_1^- = \gamma_2^+ = \gamma_2^-, \tag{14.4}$$

$$\frac{1}{c_1}\left(\gamma_1^{+'} + \frac{1}{\rho_1}\gamma_1^+\right) = \frac{1}{c_2}\left(\gamma_2^{+'} + \frac{1}{\rho_1}\gamma_2^+\right) = \frac{1}{c_1}\left(-\gamma_1^{-'} - \frac{1}{\rho_1}\gamma_1^-\right) = \frac{1}{c_2}\left(-\gamma_2^{-'} - \frac{1}{\rho_1}\gamma_2^-\right). \tag{15.4}$$

Conditions (14.4) imply that all four rays pass through a single point on the boundary (11.4), relations (15.4) are equivalent to the applicability of Snell's law for reflection of the rays at the boundary (11.4), and in reflection the "angle of incidence is equal to the angle of reflection." Continuity of the field produces the equation

$$\begin{gathered} D_1^- \frac{1}{\sqrt{\gamma_1^-}} e^{-\frac{i}{2}\frac{\gamma_1^{-'}}{\gamma_1^-} K_1 x^2 - iK_1 y} + D_1^+ \frac{1}{\sqrt{\gamma_1^+}} e^{\frac{1}{2}\frac{\gamma_1^{+'}}{\gamma_1^+} K_1 x^2 + iK_1 y} = \\ = C_1^- \frac{1}{\sqrt{\gamma_2^-}} e^{-\frac{i}{2}\frac{\gamma_2^{-'}}{\gamma_2^-} K_2 x^2 - iK_2 y} + C_1^+ \frac{1}{\sqrt{\gamma_2^+}} e^{\frac{i}{2}\frac{\gamma_2^{+'}}{\gamma_2^+} K_2 x^2 + iK_2 y}, \end{gathered} \tag{16.4}$$

and continuity of the normal derivative of the field (up to principal terms) is equivalent to the relation

$$D_1^- \frac{(-iK_1)}{\sqrt{\gamma_1^-}} e^{-\frac{i}{2}\frac{\gamma_1^{-'}}{\gamma_1^-} K_1 x^2 - iK_1 y} + D_1^+ \frac{iK_1}{\sqrt{\gamma_1^+}} e^{\frac{i}{2}\frac{\gamma_1^{+'}}{\gamma_1^+} K_1 x^2 + iK_1 y} =$$

$$= C_1^- \frac{(-i\kappa_2)}{\sqrt{\gamma_2^-}} e^{-\frac{i}{2}\frac{\gamma_2^{-\prime}}{\gamma_2^-}\kappa_2 x^2 - i\kappa_2 y} + C_1^+ \frac{i\kappa_2}{\sqrt{\gamma_2^+}} e^{\frac{i}{2}\frac{\gamma_2^{+\prime}}{\gamma_2^+}\kappa_2 x^2 + i\kappa_2 y},$$

$$\left(\kappa_j = \frac{\omega}{c_j}\right). \qquad (17.4)$$

Here the coefficients $\mathcal{D}_1^{\pm}$ characterize the wave in the first medium, and $C_1^{\pm}$ characterize it in the second medium. Using Eqs. (14.4) and (15.4), we readily obtain the equations

$$\begin{gathered} \mathcal{D}_1^- e^{-i\kappa_1 l_1} + \mathcal{D}_1^+ e^{i\kappa_1 l_1} = C_1^- e^{-i\kappa_2 l_1} + C_1^+ e^{i\kappa_2 l}, \\ \mathcal{D}_1^-(-i\kappa_1) e^{-i\kappa_1 l_1} + \mathcal{D}_1^+ i\kappa_1 e^{i\kappa_1 l_1} = C_1^-(-i\kappa_2) e^{-i\kappa_2 l_1} + C_1^+ i\kappa_2 e^{i\kappa_1 l_1}. \end{gathered} \qquad (18.4)$$

Moreover, the wave characterized in the first medium by $\mathcal{D}_1^{\pm}$ and the wave characterized in the first medium by $C_0^{\pm}$ are identical, and a complex ray $x = \gamma_1^+(y)$ incident on the boundary (11.4) is transformed into a ray characterized by the same vector as the ray incident on the boundary $y = l_0 + \frac{1}{2\rho_0} x^2$, up to the Floquet factor $e^{i\varphi_1}$. In order for the indicated waves to be identical, clearly, the following equations must be fulfilled:

$$\begin{gathered} \mathcal{D}_1^- = e^{-i\frac{\varphi_1}{2}} C_0^-, \\ \mathcal{D}_1^+ = C_0^+. \end{gathered} \qquad (19.4)$$

We deduce the following equations analogously:

$$\left.\begin{gathered} \mathcal{D}_2^- e^{-i\kappa_2 l_2} + \mathcal{D}_2^+ e^{i\kappa_2 l_2} = C_2^- e^{-i\kappa_3 l_2} + C_2^+ e^{i\kappa_3 l_2} \\ \mathcal{D}_2^-(-i\kappa_2) e^{-i\kappa_2 l_2} + \mathcal{D}_2^+ i\kappa_2 e^{i\kappa_2 l_2} = C_2^-(-i\kappa_3) e^{-i\kappa_3 l_2} + C_1^+ i\kappa_3 e^{i\kappa_3 l_2}, \end{gathered}\right\} \qquad (20.4)$$

$$\left.\begin{gathered} \mathcal{D}_2^- = e^{-i\frac{\varphi_1}{2}} C_1^- \\ \mathcal{D}_2^+ = C_1^+, \end{gathered}\right\} \qquad (21.4)$$

$$\mathcal{D}_3^- e^{-i\kappa_3 l_3} + \mathcal{D}_3^+ e^{i\kappa_3 l_3} = 0, \qquad (22.4)$$

$$\left.\begin{gathered} \mathcal{D}_3^- = e^{-i\frac{p_3}{2}} C_2^- \\ \mathcal{D}_3^+ = C_2^+. \end{gathered}\right\} \qquad (23.4)$$

Equations (10.4) and (18.4)-(23.4) for a homogeneous system of linear algebraic equations having a nontrivial solution. From this we could derive an equation for the frequencies, but shall proceed to do so at once for a more general case.

§ 5. Higher Modes and the Frequency Equation

We mention first of all the simple matter of deducing from the fulfillment of conditions analogous to (8.4), (14.4), and (15.4) at every reflecting and refracting boundary the following relations, which hold at the boundary $y = \ell_j + \frac{1}{2\rho_j} x^2$:

$$\frac{\sqrt{\pm W(\mathrm{Re}\,\gamma_j^{\pm}, \mathrm{Im}\,\gamma_j^{\pm})}}{\sqrt{c_j}\;|\gamma_j^{\pm}|} = \frac{\sqrt{\pm W(\mathrm{Re}\,\gamma_{j+1}^{\pm}, \mathrm{Im}\,\gamma_{j+1}^{\pm})}}{\sqrt{c_{j+1}}\;|\gamma_{j+1}^{\pm}|}. \tag{1.5}$$

In (1.5) $W(\ldots)$ denotes the Wronskian of the corresponding functions. The expressions under the radical are positive [see Eq. (4.4) and the discussion following it].

The ensuing constructions rely on the fact that the operator*

$$\frac{1}{c_j} \zeta(y) \frac{1}{\sqrt{\omega}} \frac{\partial}{\partial x} \mp \zeta'(y) \sqrt{\omega}\, x \tag{2.5}$$

commutes with the differential operator on the left-hand side of Eq. (3.4) [the choice of signs in (3.4) and (2.5) is opposite to $\zeta(y) = f y + g$, where f, $g = \mathrm{const}$ are arbitrary]. For the j-th medium we interpret in the role of $\zeta(y)$ in Eq. (2.5) the complex conjugate $\gamma_j^*(y)$ of the Floquet solution. Applying the operator

$$\Lambda_j^{\pm} = \frac{1}{c_j} \gamma_j^*(y) \frac{1}{\sqrt{\omega}} \frac{\partial}{\partial x} \mp \gamma_j^{*\prime} \sqrt{\omega}\, x$$

q times, we obtain the following solution of the parabolic equations:

$$(\Lambda_j^{\pm})^q \frac{1}{\sqrt{\gamma_j^{\pm}}} e^{\pm \frac{i}{2} \frac{\gamma_j^{\pm\prime}}{\gamma_j^{\pm}} \kappa_j x^2} = \left(\frac{\sqrt{\pm W(\mathrm{Re}\,\gamma_j^{\pm}, \mathrm{Im}\,\gamma_j^{\pm})}}{\sqrt{c_j}\;|\gamma_j^{\pm}|} \right)^q \times$$

$$\times \left(\gamma_j^{*\pm}\right)^q \frac{1}{\sqrt{\gamma_j^{\pm}}} e^{\pm \frac{i}{2} \frac{\gamma_j^{\pm\prime}}{\gamma_j^{\pm}} \kappa_j x^2} H_q\left(\frac{\sqrt{\pm W(\mathrm{Re}\,\gamma_j^{\pm}, \mathrm{Im}\,\gamma_j^{\pm})}}{|\gamma_j^{\pm}|} \sqrt{\kappa_j}\, x \right), \tag{3.5}$$

$$\kappa_j = \frac{\omega}{c_j}.$$

Here H_q is the q-th Hermite polynomial. Equation (3.5) is quickly justified by induction. We multiply the functions (3.5) by $\exp(\pm i \kappa_j y)$ and the constants $C_j^{\pm}$ or $\mathcal{D}_j^{\pm}$ [as previously the function (4.4)] and require that the boundary conditions be met.

It is to be noted that through Eqs. (8.4), (14.4), (15.4), and (1.5) we arrive at the same equations (10.4), (18.4), (20.4), and (22.4). Equations (19.4), (21.4), and (23.4) are replaced in this case by the equations

$$\mathcal{D}_j^- = e^{-i\left(\frac{1}{2}+q\right)\varphi_j} C_{j-1}^-, \qquad \mathcal{D}_j^+ = C_{j-1}^+. \tag{4.5}$$

*The method used to derive operators of the type (2.5) is given in [8]. Some very opportune applications of these operators in the theory of open resonators are found in [9].

Relations (4.5) are obtained in the same way as relations (19.4), (21.4), and (23.4); it is important to note that the complex "ray" $x = \gamma_j^*(y)$ traveling "downward," on being reflected first from the $(j-1)$ st boundary and then from the j-th boundary, is transformed into the ray $x = e^{-i\varphi_j}\gamma_j^*(y)$, while the function $(\gamma_j^*)^q$ clearly acquires the multiplier $\exp(-i\varphi_j q)$.

Now the frequency equation may be written in matrix form at once for arbitrary $q = 0, 1, 2, \ldots$. It follows directly from Eq. (10.4) that, up to an insignificant factor,

$$C_0^- = ie^{i\kappa_1 l_0}; \quad C_0^+ = -ie^{-i\kappa_1 l_0}. \tag{5.5}$$

We then use the latter expressions to find successively $\mathfrak{D}_1^\pm$, $C_1^\pm$, $\mathfrak{D}_2^\pm$, $C_2^\pm$, and $\mathfrak{D}_3^\pm$. The coefficients $\mathfrak{D}_3^\pm$, however, must satisfy Eq. (22.4), the latter requirement yielding an equation that must be satisfied by the frequency ω. This equation has the form

$$\left(e^{-i\kappa_3 l_3}, e^{i\kappa_3 l_3}\right) E_3^{(q)} C_3^{-1} \mathfrak{F}_2 E_2^{(q)} C_2^{-1} \mathfrak{F}_1 E_1^{(q)} \begin{pmatrix} ie^{i\kappa_1 l_0} \\ -ie^{-i\kappa_1 l_0} \end{pmatrix} = 0. \tag{6.5}$$

Here

$$E_j^{(q)} = \begin{pmatrix} e^{-i\varphi_j (q+\frac{1}{2})} & 0 \\ 0 & 1 \end{pmatrix}, \tag{7.5}$$

$$\mathfrak{F}_j = \begin{pmatrix} e^{-i\kappa_j l_j}, & e^{i\kappa_j l_j} \\ -i\kappa_j e^{-i\kappa_j l_j}, & i\kappa_j e^{i\kappa_j l_j} \end{pmatrix}, \tag{8.5}$$

$$C_1 = \begin{pmatrix} e^{-i\kappa_j l_j}, & e^{i\kappa_j l_j} \\ -i\kappa_j e^{-i\kappa_j l_j}, & i\kappa_j e^{i\kappa_j l_j} \end{pmatrix}. \tag{9.5}$$

Multiplying the one-row matrix $(\exp(-i\kappa_3 l_3), \exp i\kappa_3 l_3)$ by the vector $E_3^{(q)} C_3^{-1} \ldots \begin{pmatrix} ie^{-i\kappa_1 l_0} \\ -ie^{-i\kappa_1 l_0} \end{pmatrix}$, the result is clearly no longer a vector or a matrix, but a number which, when set equal to zero, brings us to an equation for ω. It has the form

$$\begin{aligned} &(\kappa_1+\kappa_2)(\kappa_2+\kappa_3)\sin(\nu_1+\nu_2+\nu_3) + (\kappa_2-\kappa_1)(\kappa_2+\kappa_3)\sin(\nu_1-\nu_2-\nu_3) + \\ &+ (\kappa_2-\kappa_1)(\kappa_3-\kappa_1)\sin(\nu_1-\nu_2+\nu_3) + (\kappa_1+\kappa_2)(\kappa_3-\kappa_2)\sin(\nu_1+\nu_2-\nu_3) = 0, \end{aligned} \tag{10.5}$$

where

$$\nu_j = \frac{\varphi_j}{2}\left(\frac{1}{2}+q\right) + \kappa_j h_j; \quad h_j = l_j - l_{j-1}; \quad \kappa_j = \frac{\omega}{c_j}. \tag{11.5}$$

The expressions for φ_j are well known (see, e.g., [1, 2]). In our notation*

$$\varphi_j = 2\left[\arccos\sqrt{\left(1-\frac{\varrho_{j-1}}{h_j}\right)\left(1+\frac{\varrho_j}{h_j}\right)}\,\Lambda_j + \frac{\pi}{2}(1-\Lambda_j)\right], \tag{12.5}$$

*In order that expression (12.5) be real it is necessary and sufficient that the expression under the radical lie between zero and one. This condition is tantamount to stability of the j-th partial resonator.

$$\Lambda_j = \text{sign}\left(\frac{2h_j}{\rho_j \rho_{j-1}} - \frac{1}{\rho_j} + \frac{1}{\rho_{j-1}}\right)$$

We shall omit the voluminous computations by which Eqs. (10.5)-(12.5) are derived.

§ 6. Proof of the Coincidence of $|\gamma_i^+(y)|$ and $|\gamma_j^-(y)|$; Relationship between Resonator Stability and the Solvability of the Boundary-Value Problem for the Parabolic Equation

We now show that the analytical expressions obtained in §§ 4 and 5 for the eigenfunctions are the same as the expressions obtained by the more conventional approach to the boundary-value problem for the parabolic equation.

First let the Floquet exponent for the j-th partial resonator $\varphi_j \neq$ an integral multiple of π. In this case $|\gamma_j^+(y)| = |\gamma_j^-(y)|$.

Thus, for the eigenvector $\vec{\xi}_j^-$ [$\vec{\xi}_1 = \vec{\xi}_1^- = \vec{c}_1 + i\vec{c}_2$, $\vec{\xi}_2 = B_1 \vec{\xi}_1$, $\vec{\xi}_3 = B_2 B_1 \xi_1$ (see § 3)] of the matrix $\Gamma_j \Gamma_{j-1}$ the following equation holds:

$$\Gamma_j \Gamma_{j-1} \vec{\xi}_j^- = e^{i\varphi_j} \xi_j^-,$$

from which we obtain, using the fact that $\Gamma_j = \Gamma_j^{-1}$; $\Gamma_{j-1} = \Gamma_{j-1}^{-1}$,

$$\Gamma_j \Gamma_{j-1} (\Gamma_{j-1} \xi_j^-) = e^{-i\varphi_j} (\Gamma_{j-1} \xi_j^-).$$

Inasmuch as $\exp(\pm i\varphi_j) = \pm 1$, the eigenvalues $\exp(\pm i\varphi_j)$ are simple, and

$$\Gamma_{j-1} \vec{\xi}_j^- = \mu \vec{\xi}_j^{-*} \tag{1.6}$$

($\mu = |\mu| e^{i\theta}$, and the components of the vector $\vec{\xi}_j^{-*}$ and the components of the vector $\vec{\xi}_j^-$ are complex conjugate numbers). Let L_j be a 2×2 matrix whose first column is equal to $\text{Re}\,\vec{\xi}_j$ and whose second column is equal to $\text{Im}\,\vec{\xi}_j$ ($L_1 = C$; $L_2 = B_1 C$; $L_3 = B_2 B_1 C$; see § 3).

Equation (1.6) may how be rewritten in the form

$$L_{j-1} L_j = (\text{Re}\,\mu \vec{\xi}_j^-, \text{Im}\,\mu \vec{\xi}_j^-) = |\mu| (\text{Re}\,\vec{\xi}_j^- \cos\theta + \text{Im}\,\vec{\xi}_j^- \sin\theta,$$

$$\text{Re}\,\vec{\xi}_j^- \sin\theta - \text{Im}\,\vec{\xi}_j^- \cos\theta) = |\mu| (\text{Re}\,\vec{\xi}_j, \text{Im}\,\vec{\xi}_j) \cdot \begin{pmatrix} \cos\theta, & \sin\theta \\ -\sin\theta, & \cos\theta \end{pmatrix}.$$

As usual, we denote by $(\vec{a}, \vec{b})$ the 2×2 matrix whose first (second) column is the vector $\vec{a}$ ($\vec{b}$). Going to the determinants and invoking the fact that $\det L_j \neq 0$ (otherwise the vectors $\vec{\xi}_j^-$ and $\vec{\xi}_j^{-*}$ corresponding to different eigenvalues would be linearly dependent), we obtain $|\mu|^2 = 1$, i.e., $\mu = e^{i\theta}$

The coincidence of $|\gamma_j^+(y)|$ and $|\gamma_j^-(y)|$ follows readily from the relations

$$\gamma_j^\pm(y) = \alpha^\pm y + \beta^\pm; \quad \begin{pmatrix} \alpha^- \\ \beta^- \end{pmatrix} = \vec{\xi}^-; \quad \begin{pmatrix} \alpha^+ \\ \beta^+ \end{pmatrix} = \Gamma_0 \vec{\xi}; \quad \Gamma_0 \vec{\xi}^- = e^{i\theta} \vec{\xi}^{-*}.$$

If the j-th partial resonator is stable but the inequality $e^{i\varphi_j} \neq \pm 1$ is not true, the resonator is called a confocal resonator. In this case it is a simple matter from the calculations to obtain the well-known formula $e^{\pm i\varphi_j} = -1$, $\rho_{j-1} = -\rho_j = \ell_j - \ell_{j-1}$. If a confocal resonator is a partial resonator of a stable multilayered resonator, it can terminate only in a nonconfocal resonator (this follows at once from geometrical considerations).

At the boundary of a confocal with a noncofocal resonator the boundary conditions (14.4) and (15.4) (with 1 replaced by j and 2 by $j+1$) must be met for the corresponding Floquet solutions. Let the j-th resonator be confocal and the $(j+1)$st nonconfocal. For the nonconfocal component $|\gamma^+_{j+1}| \equiv |\gamma^-_{j+1}|$, which in combination with the equation

$$\operatorname{Re}\left(-\frac{\gamma^{-\prime}_{j+1}}{\gamma^-_{j+1}} - \frac{1}{\rho_j}\right) = \operatorname{Re}\left(\frac{\gamma^{+\prime}_j}{\gamma^+_j} + \frac{1}{\rho_j}\right)$$

implies that

$$\frac{|\gamma^\pm_{j+1}|'}{|\gamma^\pm_{j+1}|} - \frac{1}{\rho_j} = 0.$$

Turning once again to Eqs. (14.4) and (15.4), we obtain for the confocal resonator at the point $y = \ell_j$

$$|\gamma^+_j(y)| = |\gamma^-_j(y)| \;\left(= |\gamma^+_{j+1}(y)| = |\gamma^-_{j+1}(y)|\right),$$

$$\frac{|\gamma^\pm_j(y)|'}{|\gamma^\pm_j(y)|} + \frac{1}{\rho_j} = \frac{c_j}{c_{j+1}}, \qquad \frac{|\gamma^\pm_{j+1}|'}{|\gamma^\pm_{j+1}|} + \frac{1}{\rho_j} = 0.$$

Consequently, at $y = \ell_j$ we have coincidence between $|\gamma^+_j|$ and $|\gamma^-_j|$, as well as between $|\gamma^+_j|'$ and $|\gamma^-_j|'$. From this and from the coincidence at $y = \ell_j$ of

$$\operatorname{Im}\frac{\gamma^{-\prime}_j}{\gamma^-_j} = -\operatorname{Im}\frac{\gamma^{+\prime}_j}{\gamma^+_j}$$

the identity $|\gamma^+_j(y)| = |\gamma^-_j(y)|$ is easily deduced algebraically.

Setting $|\gamma^\pm_j(y)| = \theta_j(y)$ (in both the confocal and the nonconfocal case) and making use of Eq. (3.5), we arrive at the following expression for waves propagating in the j-th partial resonator:

$$(\Lambda^\pm_j)^q \frac{1}{\sqrt{\gamma^\pm_j}}\, e^{\pm\frac{i}{2}\frac{\gamma^{\pm\prime}_j}{\gamma^\pm_j}K_j x^2} = \frac{1}{\sqrt{\theta_j}}\frac{1}{\sqrt[4]{c_j}}\, e^{\pm\frac{i}{2}\frac{\theta'_j}{\theta_j}K_j x^2 - \frac{1}{2\theta_j^2}K_j x^2 \pm (q+\frac{1}{2})i\int_{\ell_{j-1}}^{y}\frac{dy}{\theta_j^2(y)}} H_q\left(\frac{\sqrt{K_j}\,x}{\theta_j}\right). \tag{2.6}$$

Here H_ν is an Hermite polynomial, and θ solves the nonlinear boundary-value problem

$$\theta''\theta^3 = 1; \quad \left.\frac{\theta'}{\theta}\right|_{y=\ell_{j-1}} + \frac{1}{\rho_{j-1}} = 0; \quad \left.\frac{\theta'}{\theta}\right|_{y=\ell_j} + \frac{1}{\rho_j} = 0.$$

Calculations show that

$$\theta_j = \sqrt{\frac{|a(y-\ell_j)^2 + 2b(y-\ell_j) + c|}{\sqrt{D}}};$$

$$a = \frac{1}{\rho_{j-1}} - \frac{1}{\rho_j} + \frac{2h_j}{\rho_{j-1}\rho_j}, \quad b = -\frac{1}{\rho_{j-1}} h_j \left(H\frac{h_j}{\rho_j}\right), \quad c = h_j\left(H\frac{h_j}{\rho_j}\right), \tag{3.6}$$

$$D = ac - b^2 = \frac{h_j}{\rho_j^2 \rho_{j-1}^2}(h_j + \rho_j)(h_j - \rho_{j-1})(\rho_{j-1} - \rho_j - h_j);$$

$$\int_{\ell_{j-1}}^{y} \frac{dy}{\theta_j^2(y)} = \tan^{-1}\frac{|a|(y-\ell_{j-1}) - \frac{1}{\rho}|c|}{\sqrt{D}} + \tan^{-1}\frac{|c'|}{\rho_j\sqrt{D}}. \tag{4.6}$$

If in the parabolic equations (3.4), after a number of suitable transformations, we separate variables, we arrive at formulas of the type (2.6). Introducing

$$\gamma_j^{+} = \theta_j \exp \pm i\int_{\ell_0} \frac{dy}{\theta_j^2(y)} \cdot \chi_j^{\pm} \qquad (\chi_j^{\pm} = \text{const})$$

and requiring that the boundary conditions be met at the reflecting and refracting boundaries, we come to the conclusion that, given a proper choice of $\chi^{\pm} = \text{const}$, $\gamma_j(y)$ will represent the Floquet solutions for each partial resonator, matched as in § 3. The existence of these matched Floquet solutions implies stability of the multilayered resonator. The attendant computations are elementary, and we shall not go through them here.

§ 7. Resonator Stability and the "Compatibility Conditions" of Boitsov

It follows from the arguments of the preceding section that at every reflecting or refracting boundary

$$\mathrm{Re}\left(\frac{\gamma_j'}{\gamma_j} + \frac{1}{\rho_j}\right) = 0. \tag{1.7}$$

In order for the boundary-value problem to be solvable for the parabolic equation it is necessary and sufficient that the following equations hold:

$$\pm\frac{1}{c_j}\left(\left.\frac{\gamma_j^{\pm\prime}}{\gamma_j^{\pm}}\right|_{y=\ell_j} + \frac{1}{\rho_j}\right) = \frac{\pm 1}{c_{j+1}}\left(\left.\frac{\gamma_{j+1}^{\pm\prime}}{\gamma_{j+1}^{\pm}}\right|_{y=\ell_j} + \frac{1}{\rho_{j+1}}\right), \tag{2.7}$$

which by virtue of (1.7) and the fact that $|\gamma_j^{+}(y)| \equiv |\gamma_j^{-}(y)|$ are reduced to the equation

$$\frac{c_j |\gamma_j^{\pm}(y)|}{\pm W(\operatorname{Re}\gamma_j^{\pm}, \operatorname{Im}\gamma_j^{\pm})} = \frac{c_{j+1} |\gamma_j^{\pm}(y)|}{\pm W(\operatorname{Re}\gamma_{j+1}^{\pm}, \operatorname{Im}\gamma_j^{\pm})}\Bigg|_{y=\ell_j}$$

(where W is the Wronskian). Since (see § 6)

$$\gamma_j^{+}(y) = \overline{\gamma_j^{-}(y)}\, e^{i\theta}, \qquad 0 \leq \theta < 2\pi$$

(the prime denotes the complex conjugate), the above equations are reduced to the relation

$$\frac{c_j |\gamma_j^{+}(y)|}{W(\operatorname{Re}\gamma_j^{+}, \operatorname{Im}\gamma_j^{+})} = \frac{c_{j+1} |\gamma_{j+1}^{+}(y)|}{W(\operatorname{Re}\gamma_{j+1}^{+}, \operatorname{Im}\gamma_j^{+})}. \tag{3.7}$$

We note further that

$$|\gamma_j| = \sqrt{a_j y^2 + 2b_j y + c_j}, \tag{4.7}$$

$$\left.\begin{aligned} &\frac{d}{dy}|\gamma_j| + \frac{1}{\rho_j}|\gamma_j|\Big|_{y=\ell_j} = 0;\\ &\frac{d}{dy}|\gamma_j| + \frac{1}{\rho_{j-1}}|\gamma_j|\Big|_{y=\ell_{j-1}} = 0. \end{aligned}\right\} \tag{5.7}$$

Equations (4.7) and (5.7) determine $|\gamma_j|$ to a constant factor. Relation (3.7) may be rewritten in the form

$$\frac{c_j |\gamma_j|_{y=\ell_j}}{\sqrt{a_j c_j - b_j^2}} = \frac{c_{j+1} |\gamma_{j+1}|_{y=\ell_{j+1}}}{\sqrt{a_{j+1} c_{j+1} - b_{j+1}^2}}. \tag{6.7}$$

After some rather tedious computations Eq. (6.7) acquires the form

$$\sqrt{\frac{1 - \dfrac{\rho_{j-1}}{h_j}}{\left(1 + \dfrac{\rho_j}{h_j}\right)\left(\dfrac{\rho_{j-1}}{h_j} - \dfrac{\rho_j}{h_j} - 1\right)}} = c_{j+1}\sqrt{\frac{1 + \dfrac{\rho_{j+1}}{h_j + 1}}{\left(1 - \dfrac{\rho_j}{h_{j+1}}\right)\left(\dfrac{\rho_j}{h_{j+1}} - \dfrac{\rho_{j+1}}{h_{j+1}} - 1\right)}}, \tag{7.7}$$

$$h_j = \ell_j - \ell_{j-1} \qquad (j = 1, 2).$$

Conditions (7.7) are precisely the compatibility conditions of V. F. Boitsov (see [3]).* Consequently, in order for our constructions to be admissible there must be two relations between the parameters defining the resonator.

We observe that stability of each partial resonator and fulfillment of the compatibility conditions are sufficient for stability in a small multilayered resonator as a whole.

*Note that in Eqs. (8.14) of Boitsov's dissertation [3] not all the signs are given properly. If we insert suitable corrections into the formulas, the resulting relations will exactly coincide with our Eqs. (7.7).

§ 8. On the Real Character of the Roots of the Frequency Equation

The equation derived in § 5 for the frequencies has an infinite set of roots, all of which are real.* This conclusion evolves from the existence of a self-adjoint operator with a discret spectrum, whose eigenvalues are found from Eq. (6.5) or Eq. (10.5). Let us consider the following system of differential equations:

$$\frac{1}{i}\frac{\partial W_j^+}{\partial y} = \frac{\omega}{c_j} W_j^+;$$

$$-\frac{1}{i}\frac{\partial W_j^-}{\partial y} - \frac{\varphi_j\left(\frac{1}{2}+q\right)}{h_j} W_j^- = \frac{\omega}{c_j} W_j^-; \tag{1.8}$$

$$\ell_{j-1} \leq y \leq \ell_j;\quad h_j = \ell_j - \ell_{j-1};\quad j = 1, 2, 3.$$

We require that the sought-after functions $W_j^{\pm}$ satisfy the boundary conditions

$$W_0^+ + W_0^-\Big|_{y=\ell_0} = 0; \tag{2.8}$$

$$\left.\begin{aligned} W_j^+ + W_j^-\Big|_{y=\ell_j} &= W_{j+1}^+ + W_{j+1}^-\Big|_{y=\ell_j}, \\ \frac{1}{c_j}\left(W_j^+ - W_j^-\right)\Big|_{y=\ell_j} &= \frac{1}{c_{j+1}}\left(W_{j+1}^+ - W_{j+1}^-\right)\Big|_{y=\ell_j}; \end{aligned}\right\} \tag{3.8}$$

$$W_3^+ + W_3^-\Big|_{y=\ell_3} = 0. \tag{4.8}$$

The attempt to find a nonzero solution to the set of equations (1.8) subject to the boundary conditions (2.8)-(4.8) leads to Eq. (6.5) (the role of the spectral parameter is taken to ω). Thus, assuming for $\ell_{j-1} \leq y \leq \ell_j$ that

$$W_j^+ = C_j^+ e^{i\frac{\omega}{c_j}y};\qquad W_j^- = C_j^- e^{-i\frac{\omega}{c_j}y + \frac{\varphi_j}{h_j}\left(\frac{1}{2}+q\right)\left(y-\ell_{j-1}\right)} \tag{5.8}$$

or

$$W_j^+ = \mathcal{D}_{j+1}^+ e^{i\frac{\omega}{c_j}y};\qquad W_j^+ = \mathcal{D}_{j+1}^- e^{-i\frac{\omega}{c_j}y + \frac{\varphi_j}{h_j}\left(\frac{1}{2}+q\right)\left(y-\ell_j\right)}, \tag{6.8}$$

we immediately arrive at Eqs. (10.4) and (16.4)-(23.4) for $q=0$, or at Eqs. (10.4), (16.4), (18.4), (20.4), (22.4), and (4.5) for $q>0$. The problem of finding the eigenvalues and eigenfunctions of (1.8)-(4.8) is akin to the classical Sturm-Liouville problem. Its self-adjointness can be proved by construction of the Green function and the Hilbert-Schmidt theorem, following the same general procedure as in the case of the Sturm-Liouville problem (the theory of the eigenfunctions of the Sturm-Liouville problem is given, for example in the textbook [11].

*The real character of the roots is attributable to the fact that we completely ignored diffraction losses in our formulations.

LITERATURE CITED

1. Buldyrev, V. S., Shortwave asymptotic behavior of the eigenfunctions of the Helmholtz equation, Dokl. Akad. Nauk SSSR, 163(4):853-856 (1965).
2. Buldyrev, V. S., Asymptotic behavior of the eigenfunctions of the Helmholtz equation for plane convex domains, Vest. Leningrad. Univ., No. 22, pp. 38-51 (1965).
3. Boitsov, V. F., Some Problems in the Theory of Real Optical Resonators (dissertation), Leningrad. Univ. (1968).
4. Leng, S., Algebra, Mir, Moscow (1968), 564 pages.
5. Kurosh, A. G., Course in Higher Algebra, Moscow (1968).
6. Vilenkin, N. Ya., Special Functions and the Theory of Group Representations, Nauka, Moscow (1965).
7. Babich, V. M., The asymptotic behavior of "quasi-eigenvalues" of the exterior problem for the Laplace operator, in: Topics in Mathematical Physics, Vol. 2, Consultants Bureau, New York (1968), pp. 1-7.
8. Babich, V. M., Eigenfunctions concentrated in a neighborhood of a closed geodesic, Seminars in Mathematics, Vol. 9: Mathematical Problems in Wave Propagation Theory, Consultants Bureau, New York (1970), pp. 7-26.
9. Popov, M. M., Asymptotic behavior of certain subsequences of eigenvalues of boundary-value problems for the Helmholtz equation in the multidimensional case, Dokl. Akad. Nauk SSSR, 184(5):1076-1079 (1969).
10. Popov, M. M., Eigenmodes of Multiply Reflecting Resonators, Vest. Leningrad. Univ., Vol. 22, No. 4, pp. 42-54 (1969).
11. Smirnov, V. I., Course in Higher Mathematics, Vol. 4, Gostekhizdat, Moscow (1951).

CONCENTRATED ASYMPTOTIC BEHAVIOR OF THE SOLUTION TO THE SCHRÖDINGER EQUATION IN THE VICINITY OF A CLASSICAL TRAJECTORY

V. M. Babich and Yu. P. Danilov

In the present article we formulate the asymptotic behavior as $h \to 0$ of the solution to the Cauchy problem for the Schrödinger equation

$$i h \frac{\partial \Psi}{\partial t} = -\frac{h^2}{2} \Delta \Psi + U(x) \Psi$$

$$x = (x_1, x_2, \ldots, x_m), \quad t \in [0, T] \tag{1}$$

with time-independent potential $U(x)$:

$$U(x) \in C^\infty(R^m) \tag{2}$$

$$U(x) \geq \text{const} \tag{3}$$

$$|U(x)| \leq C \cdot |x|^{\kappa^*} \quad \text{for} \quad |x| > \text{const} > 0 \tag{4}$$

in the case when the initial condition

$$\Psi(x,t)\Big|_{t=0} = \Psi_0(x) \tag{5}$$

has a particular special form.

The resulting asymptotic behavior is uniform on the whole time interval $[0, T]$ of existence of a solution to the corresponding classical problem

$$\left.\begin{aligned} &\ddot{x}_\kappa = \frac{\partial U}{\partial x_\kappa}, \qquad \kappa = 1, 2, \ldots, m, \\ &x_\kappa\Big|_{t=0} = x_\kappa^0, \qquad \dot{x}_\kappa\Big|_{t=0} = \dot{x}_\kappa^0 \end{aligned}\right\} \tag{6}$$

and satisfies the condition

$$\overline{[x-X(t)]^2} = \int [x-X(t)]^2 |\Psi(x,t)|^2 dx = o(h), \tag{7}$$

where $X(t)$ is a solution to problem (6). We refer to (7) as the condition of concentration of the function $\Psi(x,t)$ in the vicinity of the classical trajectory $X(t)$.

We formulate the asymptotic behavior of the solution to the problem (1)-(5) on the basis of an analysis in the first approximation of the extremals of the action functional

$$S(x,t) = \int \left\{ \frac{1}{2} \sum_k \dot{x}_k^2 - U(x) \right\} dt, \tag{8}$$

i.e., on an investigation of classical particle trajectories close to the original path $X(t)$.

The assumptions (2) and (4) made above with regard to the smoothness and behavior at infinity of the potential $U(x)$ are sufficient for our ensuing constructions and proofs. Condition (3) guarantees self-adjointness of the operator

$$\hat{H} = -\frac{h^2}{2}\Delta + U(x)$$

(see, e.g., [1]).

Since we shall be concerned solely with the neighborhood of a fixed classical trajectory $X(t)$, it is reasonable to transform to the variables $z = (z_1, z_2, \ldots, z_m)$:

$$x_k = X_k(t) + z_k. \tag{9}$$

If we introduce the functions $\Psi(z,t)$ and $V(z,t)$:

$$\Psi(z,t) = \Psi(X(t)+z,t) = \Psi(x,t),$$

$$V(z,t) = U(X(t)+z) = U(x)$$

and recognize that

$$\frac{\partial \Psi}{\partial z_k} = \frac{\partial \Psi}{\partial x_k} \quad \text{and} \quad \frac{\partial \Psi}{\partial t} = \frac{d}{dt}\Psi(X(t)+z,t) = \sum_k \frac{\partial \Psi}{\partial x_k}\dot{X}_k + \frac{\partial \Psi}{\partial t},$$

we are entitled to write Eq. (1) as follows:

$$ih\left(\frac{\partial \Psi}{\partial t} - \sum_k \dot{X}_k \frac{\partial \Psi}{\partial z_k}\right) = -\frac{h^2}{2}\Delta\Psi + V(z,t)\Psi. \tag{10}$$

We seek the solution to Eq. (10) in the form

$$\Psi(z,t) = \Phi(z,t)\exp\left(\frac{i}{h}\right)[S_0(t) + \sum_k \dot{X}_k z_k], \tag{11}$$

where

$$S_0(t) = \int_0^t \left\{ \frac{1}{2}\sum_k \dot{X}_k^2(\tau) - U(X(\tau)) \right\} d\tau.$$

The substitution (11) differs from the usual [2] quasi-classical approximation

$$\Psi(x,t) \sim A(x,t)\, e^{\frac{i}{h} S(x,t)}$$

to the extent that the exponent of the exponential function is formed of the linear terms of the decomposition of the action $S(x,t)$ in the neighborhood of the classical particle trajectory into a power series in $z = x - X(t)$.

If now in Eq. (10) we replace the potential $V(z,t)$ by its Maclaurin series

$$\sum_{n \geq 0} \frac{1}{n!} \sum_{\substack{s_1, \ldots, s_m \\ s_1 + \ldots + s_m = n}} \frac{\partial^n V}{\partial z_1^{s_1} \ldots \partial z_m^{s_m}} \bigg|_{z=0} z_1^{s_1} \ldots z_m^{s_m},$$

for the function $\Phi(z,t)$ we obtain the formal equation

$$ih \frac{\partial \Phi}{\partial t} = -\frac{h^2}{2} \Delta \Phi + \sum_{n \geq 2} \frac{1}{n!} \sum_{\substack{s_1, \ldots, s_m \\ s_1 + \ldots + s_m = n}} \frac{\partial^n V}{\partial z_1^{s_1} \ldots \partial z_m^{s_m}} \bigg|_{z_1 = 0} z_1^{s_1} \ldots z_m^{s_m} \Phi. \tag{12}$$

After division of both sides of (12) by h and the introduction of the new variables

$$\mu_k = \frac{z_k}{h^{1/2}} \qquad k = 1, 2, 3, \ldots, m \tag{13}$$

and the notation

$$\mu^{[s]} = \mu_1^{s_1} \cdot \mu_2^{s_2} \cdot \ldots \cdot \mu_m^{s_m};$$

$$\sum_{|s| = n} \cdot \equiv \sum_{\substack{s_1, s_2, \ldots, s_m \\ s_1 + s_2 + \cdots + s_m = n}};$$

$$V_{k\ell}(t) \equiv \frac{\partial^2 V}{\partial z_k \partial z_\ell} \bigg|_{z=0} (t);$$

$$V_{[s]}(t) \equiv \frac{\partial^{|s|} V}{\partial z_1^{s_1} \ldots \partial z_m^{s_m}} \bigg|_{z=0} (t),$$

Eq. (12) assumes the form

$$i \frac{\partial U(\mu,t)}{\partial t} + \frac{1}{2} \Delta U(\mu,t) - \frac{1}{2} (V(t)\vec{\mu}, \vec{\mu}) U(\mu,t) = \\ = \sum_{n \geq 1} \frac{h^{\frac{n}{2}}}{(n+2)!} \sum_{|s| = n+2} V_{[s]}(t)\, \mu^{[s]}\, U(\mu,t), \tag{14}$$

where

$$\Delta \equiv \frac{\partial^2}{\partial \mu_1^2} + \frac{\partial^2}{\partial \mu_2^2} + \dots + \frac{\partial^2}{\partial \mu_m^2}$$

and $V(t) = \| V_{\kappa\ell}(t) \|$ is a real symmetric matrix.

If the right-hand side of (14) is set equal to zero, rather than to the formal power series in the small parameter $h^{\frac{1}{2}}$ we obtain the Schrödinger equation with square-law potential:

$$i \frac{\partial U(\mu, t)}{\partial t} + \frac{1}{2} \Delta U(\mu, t) - \frac{1}{2} (V(t)\vec{\mu}, \vec{\mu}) U(\mu, t) = 0,$$

whose solutions we formulate after [4]. It is a well-known fact that the quasi-classical approximation of the Schrödinger equation with square-law potential is an exact solution. This lays the groundwork for our subsequent formulations, as well as those of [4].

If we seek the fundamental solution $U_0(\mu, t)$:

$$U_0(\mu, t) \not\equiv 0, \quad |U_0(\mu, t)| \xrightarrow[|\mu| \to \infty]{} 0$$

of the equation

$$\mathcal{L} U(\mu, t) \equiv 2i \frac{\partial U}{\partial t} + \Delta U - (V(t)\vec{\mu}, \vec{\mu}) U = 0 \tag{15}$$

in the form

$$U_0(\mu; t) = P(t) e^{\frac{i}{2}(\Gamma(t)\vec{\mu}, \vec{\mu})},$$

then for the desired matrix $\Gamma(t) = \| \Gamma_{\kappa\ell}(t) \|$ and the function $P(t)$ we obtain the equations

$$\Gamma' + \Gamma^2 + V = 0 \tag{16}$$

and

$$P' + \frac{1}{2} \operatorname{Spur} \Gamma \cdot P = 0. \tag{17}$$

The substitution $\Gamma(t) = Y'(t) \cdot Y^{-1}(t)$ reduces the Ricatti matrix equation (16) to the linear form

$$Y'' + VY = 0, \tag{18}$$

and the solution to Eq. (17) assumes the form

$$P(t) = \frac{1}{\sqrt{\det Y}}$$

We construct a solution $Y(t)$ to (18) guaranteeing the required properties of the fundamental solution $U_0(\mu, t)$ of the equation $\mathcal{L} U = 0$, from the linearly-independent solutions of the vector equation

$$\vec{Y}'' + V \vec{Y} = 0,$$
$$\vec{Y}(t) = (Y_1(t), \dots, Y_m(t)), \quad t \in [0, T]. \tag{19}$$

Relation (19) is the set of Euler equations for the quadratic functional

$$\mathcal{J}_2 = \int_0^T \left\{ \frac{1}{2} \sum_\kappa \dot{z}_\kappa^2 - \frac{1}{2} \sum_{\kappa, \ell} V_{\kappa\ell}(t) z_\kappa z_\ell \right\} dt,$$

which is equal to the second variation of the action $S(x,t)$ in the neighborhood of the trajectory $X(t)$, i.e., (19) is the Jacobi equation for the extremal $X(t)$ of the functional (8).

The solutions to Eq. (19) represent classical particle trajectories that are close in the first approximation to the original path $X(t)$. Let us consider m solutions $\vec{Y}_1, \vec{Y}_2, \ldots, \vec{Y}_m$ of Eq. (19) satisfying the initial conditions

$$\vec{Y}_\kappa(0) = +\frac{i}{\sqrt{2}}\vec{l}_\kappa;$$

$$\vec{Y}'_\kappa(0) = -\frac{1}{\sqrt{2}}\vec{l}_\kappa, \quad \kappa = 1, 2, \ldots, m, \tag{20}$$

along with the m vectors $\vec{Y}_1^*, \vec{Y}_2^*, \ldots, \vec{Y}_m^*$ comprising the complex conjugates of the solutions $\vec{Y}_1, \vec{Y}_2, \ldots, \vec{Y}_m$, under the initial conditions

$$\vec{Y}_\tau^*(0) = -\frac{i}{\sqrt{2}}\vec{l}_\tau;$$

$$\vec{Y}_\tau^{*\prime}(0) = -\frac{1}{\sqrt{2}}\vec{l}_\tau, \quad \tau = 1, 2, \ldots, m. \tag{21}$$

The solutions $\vec{Y}_1, \ldots, \vec{Y}_m, \vec{Y}_1^*, \ldots, \vec{Y}_m^*$ chosen in accordance with conditions (20) and (21) are linearly independent, because

$$\det \left\| \begin{matrix} \vec{Y}_1 \ldots \vec{Y}_m & \vec{Y}_1^* \ldots \vec{Y}_m^* \\ \vec{Y}'_1 \ldots \vec{Y}'_m & \vec{Y}_1^{*\prime} \ldots \vec{Y}_m^{*\prime} \end{matrix} \right\|_{t=0} = \frac{1}{2^m} \left| \begin{array}{c|c} +iE & -iE \\ \hline -E & -E \end{array} \right| = (+i)^m \neq 0.$$

Moreover, for the vectors $\vec{Y}_1, \ldots, \vec{Y}_m, \vec{Y}_1^*, \ldots, \vec{Y}_m^*$ chosen in accordance with conditions (20) and (21) are linearly independent, because

$$(\vec{Y}'_\kappa, \vec{Y}_\tau^*) - (\vec{Y}_\kappa, \vec{Y}_\tau^{*\prime}) = 0, \tag{22}$$

$$(\vec{Y}'_\kappa, \vec{Y}_l) - (\vec{Y}_\kappa, \vec{Y}'_l) = i\delta_{\kappa l} \tag{23}$$

hold for all $t \in [0,T]$; they need be proved only for $t = 0$, because by Eq. (19) the left-hand sides of Eqs. (22) and (23) are time independent. The initial conditions (20) and (21) constitute one example of how independent vectors are chosen to satisfy relations (22) and (23).

If the m solutions $\vec{Y}_1, \ldots \vec{Y}_m$ are adopted as columns of the matrix $\|\vec{Y}_1 \vec{Y}_2 \cdots \vec{Y}_m\| = Y(t)$, we obtain a solution to (18) satisfying the relations

$$Y^T Y' - Y'^T Y = 0, \tag{24}$$

$$Y^+ Y' - Y'^+ Y = iE. \tag{25}$$

Also, $\det Y(t) \neq 0$ on $[0,T]$. In fact, we could deduce from $\det Y(t_1) = 0$ linear dependence of the vectors $\vec{Y}_1, \ldots, \vec{Y}_m$ at that point, i.e.,

$$\sum_{r=1}^{m} c_r \vec{Y}_r(t_1) = 0, \text{ where } \sum_{r=1}^{m} |c_r|^2 \neq 0. \tag{26}$$

On the other hand, multiplying (23) on the left by c_κ and on the right by c_ℓ^* and summing on ℓ and κ from 1 to m, we arrive at the identity

$$\left(\sum_{\kappa=1}^{m} c_\kappa \vec{Y}_\kappa'(t), \sum_{\ell=1}^{m} c_\ell \vec{Y}_\ell(t)\right) - \left(\sum_{\kappa=1}^{m} c_\kappa \vec{Y}_\kappa(t), \sum_{\ell=1}^{m} c_\ell \vec{Y}_\ell'(t)\right) = i \sum_{\kappa=1}^{m} |c_\kappa|^2,$$

which for $t=t_1$ yields $\sum_{\kappa=1}^{m} |c_\kappa|^2 = 0$, contradicting (26).

Relations (24) and (25), which are true for the formulated nonparticular solution to the matrix equation (18), ensure symmetry of the matrix $\Gamma(t)$ and positive definiteness of its imaginary part:

$$\operatorname{Im} \Gamma(t) = \tfrac{1}{2}(YY^+)^{-1}.$$

The fundamental solution to Eq. (15) assumes the form

$$u_0(\mu, t) = \frac{1}{\sqrt{\det Y}} e^{\frac{i}{2}(Y'Y^{-1}\vec{\mu}, \vec{\mu})},$$

and its modulus squared

$$|u_0(\mu, t)|^2 = \frac{1}{|\det Y|} e^{-\frac{1}{2}((YY^+)^{-1}\vec{\mu}, \vec{\mu})} \longrightarrow 0$$

as $|\vec{\mu}| \to \infty$ more rapidly than any negative power of the variables $\mu_1, \mu_2, \ldots, \mu_m$. The other solutions of the equation $\mathcal{L} U = 0$ are obtained through a consideration of the creation operators

$$\Lambda_\kappa^* = \frac{1}{i}(\vec{Y}_\kappa^*, \nabla_\mu) - (\vec{Y}_\kappa^{*\prime}, \vec{\mu}),$$

$$\left[(\vec{Y}, \nabla_\mu)\mathcal{F} = \sum_{p=1}^{m} Y_p \frac{\partial \mathcal{F}}{\partial \mu_p}, \quad \text{for } \vec{Y} = Y_1, \ldots Y_m\right]$$

and destruction operators

$$\Lambda_r = \frac{1}{i}(\vec{Y}_r, \nabla_\mu) - (\vec{Y}_r', \vec{\mu}), \qquad \kappa, r = 1, 2, \ldots, m,$$

on the basis of the solution $u_0(\mu, t)$. It can be shown by direct calculations that

$$\Lambda_r u_0(\mu, t) \equiv 0 \qquad r = 1, 2, \ldots, m$$

and that the following commutation relations are valid:

$$\Lambda_\kappa \Lambda_r - \Lambda_r \Lambda_\kappa = 0$$

$$\Lambda_\kappa^* \Lambda_r^* - \Lambda_r^* \Lambda_\kappa^* = 0$$

$$\Lambda_r \Lambda_\kappa^* - \Lambda_\kappa^* \Lambda_r = \delta_{\kappa r}.$$

Also, the operators Λ_κ^* and Λ_κ $\kappa=1,2,\dots,m$ commute with the operator $\mathcal{L}=2i\frac{\partial}{\partial t}+\Delta_\mu-(V(t)\vec{\mu},\mu)$ of Eq. (15):

$$\Lambda_\kappa^*\mathcal{L}=\mathcal{L}\Lambda_\kappa^*, \quad \Lambda_\kappa\mathcal{L}=\mathcal{L}\Lambda_\kappa.$$

This justifies the assertion that the functions

$$u_{[s]}(\mu,t)=u_{s_1,s_2,\dots,s_m}(\mu,t)=(\Lambda_1^*)^{s_1}(\Lambda_2^*)^{s_2}\cdots(\Lambda_m^*)^{s_m}u_0(\mu,t)=$$

$$=Q_{[s]}(\mu,t)\frac{1}{\sqrt{\det Y}}\,e^{\frac{i}{2}(\Gamma(t)\vec{\mu},\vec{\mu})}, \tag{27}$$

where $Q_{[s]}(\mu,t)$ is a certain polynomial of degree $|s|=\sum_{\alpha=1}^{m}s_\alpha$ with the coefficients depending on t, will be solutions to Eq. (15) for any sets $[s]=(s_1,s_2,\dots,s_m)$ of nonnegative integers s_α. For any fixed $t\in[0,T]$ the following orthogonality relations are applicable to the functions $u_{[s]}(\mu,t)$:

$$\int u_{[s']}(\mu,t)\,u^*_{[s'']}(\mu,t)\,d\mu=\begin{cases}0 & (s_1',s_2',\dots,s_m')\neq(s_1'',s_2'',\dots,s_m'')\\ s_1!\,s_2!\dots s_m!\int|u_0(\mu,t)|^2d\mu=[s]!\,(2\pi)^{\frac{m}{2}}.\end{cases}$$

The latter implies that the polynomials $Q_{[s]}(\mu,t)$ defined by Eq. (27) are linearly independent. Since there are as many of them as there are monomials $\mu^{[s]}=\mu_1^{s_1},\dots,\mu_m^{s_m}$, the polynomials $Q_{[s]}(\mu,t)$, $0\le|s|\le\varkappa$ form a basis in the finite-dimensional linear space of polynomials of $\mu_1,\mu_2,\dots,\mu_m$ of degree no higher than $\varkappa\geqslant0$.

Consequently, the system of functions $\{u_{[s]}(\mu,t)\}$ turns out for any fixed $t\in[0,T]$ to be not only orthogonal, but also complete in $L_2(R^m)$. We seek the solution to the equation

$$\mathcal{L}u(\mu,t)=2i\frac{\partial u}{\partial t}+\Delta_\mu u-(V(t)\vec{\mu},\vec{\mu})u=\sum_{n\geqslant1}\frac{2h^{\frac{n}{2}}}{(n+2)!}\sum_{|s|=n+2}V_{[s]}(t)\,\mu^{[s]}u \tag{14}$$

in the form of a formal power series in $h^{\frac{1}{2}}$:

$$u(\mu,t)=u_0(\mu,t)+\sum_{\ell\geqslant1}h^{\frac{\ell}{2}}u_\ell(\mu,t). \tag{28}$$

We call the segment

$$\Phi_{N+1}(\mu,t)=u_0(\mu,t)+\sum_{\ell=1}^{N}h^{\frac{\ell}{2}}u_\ell(\mu,t)$$

of the series (28) the N-th approximation for the solution to Eq. (14).

Substituting the series (28) into Eq. (14) and equating coefficients of like powers of $h^{\frac{1}{2}}$, we obtain a recursive system of equations for the functions $u_\ell(\mu,t)$:

$$\mathcal{L}u_1(\mu,t)=\frac{2}{3!}\sum_{|s|=3}V_{[s]}(t)\,\mu^{[s]}u_0(\mu,t)$$

$$\dots\dots\dots\dots\dots\dots\dots$$

$$\mathcal{L} U_\ell(\mu,t)=\sum_{\kappa=1}^{\ell} \frac{2}{(\kappa+2)!} \sum_{|s|=\kappa+2} V_{[s]}(t)\, \mu^{[s]} U_{\ell-\kappa}(\mu,t). \tag{29}$$

The successive solution of Eqs. (29) reveals that the functions $U_\ell(\mu, t)$ have the form of a linear combination of functions $U_{[p]}(\mu, t)$, the modulus of whose multiindex is not greater than

$$U_\ell(\mu,t)=\sum_{|p|=0}^{3\ell} C_{[p]}^{(\ell)}(t)\, U_{[p]}(\mu,t)=\Big[\sum_{|p|=0}^{3\ell} C_{[p]}^{(\ell)}(t)\, Q_{[p]}(\mu,t)\Big]\cdot U_0(\mu,t).$$

The coefficients of this linear combination are determined in terms of the coefficients $\lambda_{[q]}^{(\ell)}(t)$ of the decomposition of the right-hand side of Eq. (29) into a system of linearly-independent polynomials $Q_{[q]}(\mu,t)$, $0\le|q|\le 3\ell$, $|q|=\sum_{j=1}^{m} q_j$, multiplied by $U_0(\mu,t)$:

$$C_{[p]}^{(\ell)}(t)=-\frac{i}{2}\int_0^t \lambda_{[p]}^{(\ell)}(\tau)\, d\tau+\theta_{[p]}^{(\ell)}.$$

The constants of integration $\theta_{[p]}^{(\ell)}$ in the latter formula determine the value of the N-th approximation at $t = 0$:

$$\Phi_{N+1}(\mu,t)\Big|_{t=0} \equiv \Phi_0^{(N+1)}(\mu)=\Big[1+\sum_{\ell=1}^{N} h^{\frac{\ell}{2}} \sum_{|p|=0}^{3\ell} \theta_{[p]}^{(\ell)}\, Q_{[p]}(\mu,0)\Big]\cdot U_0(\mu,0),$$

where

$$\Phi_0^{(1)}(\mu)=U_0(\mu,0)=\left(\frac{\sqrt{2}}{i}\right)^{\frac{m}{2}} e^{-\frac{1}{2}(\mu_1^2+\mu_2^2+\dots+\mu_m^2)}.$$

By the fact that at $t = 0$ the creation operators have the form

$$\Lambda_\kappa^* = \frac{1}{\sqrt{2}}\left(\mu_\kappa-\frac{\partial}{\partial \mu_\kappa}\right)$$

and

$$\left(\mu_\kappa-\frac{\partial}{\partial\mu_\kappa}\right)^{s_\kappa} e^{-\frac{1}{2}(\mu_1^2+\dots+\mu_m^2)} = H_{s_\kappa}(\mu_\kappa)\, e^{-\frac{1}{2}(\mu_1^2+\dots+\mu_m^2)},$$

then with respect to each of the variables μ the function $Q_{[p]}(\mu,0)$ turns out to be a Hermite polynomial of order p_α:

$$Q_{[p]}(\mu,0)=\prod_{\alpha=1}^{m} 2^{-\frac{p_\alpha}{2}} H_{p_\alpha}(\mu_\alpha).$$

We set the functions $\Phi_{N+1}(\mu,t)$ constructed above in correspondence with functions executing the following operations in sequence:

1) the substitution $\rho_\kappa = \frac{z_\kappa}{h^{1/2}}$;

2) multiplication by the function

$$\exp\left(\frac{i}{h}\right)[S_0(t)+\sum_\kappa \dot{X}_\kappa z_\kappa];$$

3) transition from the variables $\vec{z}$, t to the variables $\vec{x}$, t:

$$z_\kappa = x_\kappa - X_\kappa(t), \qquad \kappa = 1, 2, \ldots, m$$

by the following scheme:

$$\Phi_{N+1}(\rho,t) \xrightarrow{①} \Phi_{N+1}(z,t) \xrightarrow{②} \Psi_{N+1}(z,t) \xrightarrow{③} \Psi_{N+1}(x,t).$$

Computation of the square of the norm of $\Psi_{N+1}(x,t)$ yields

$$\|\Psi_{N+1}(x,t)\|^2 = (2\pi h)^{\frac{m}{2}}\,[1+W_N(t,h)],$$

where

$$W_N(t,h) \leq \text{const}\cdot h^{\frac{1}{2}} \tag{30}$$

and $W_N(t,h)=0$ for $N=0$. The uniformity of the estimate (30) on $t\in[0,T]$ permits the following to be adopted as the "normalized" functions:

$$\overset{\circ}{\Psi}_{N+1}(x,t) \equiv \frac{\Psi_{N+1}(x,t)}{(2\pi h)^{\frac{m}{4}}}.$$

It can be shown by fairly straightforward, though tedious, estimates that for the solution of the Cauchy problem

$$\left.\begin{aligned} &ih\frac{\partial\Psi}{\partial t} = -\frac{h^2}{2}\Delta\Psi + U(x)\Psi, \\ &\Psi(x,t)\Big|_{t=0} = \overset{\circ}{\Psi}_{N+1}(x,0) \end{aligned}\right\} \tag{31}$$

the following estimate is valid:

$$\|\Psi(x,t) - \overset{\circ}{\Psi}_{N+1}(x,t)\| = O\left(h^{\frac{N+1}{2}}\right),$$

$$N = 0, 1, 2, \ldots,$$

which is uniform on $t\in[0,T]$.

The mean-square deviation from the classical trajectory $X(t)$ in the state $\overset{\circ}{\Psi}_{N+1}(x,t)$ is equal to

$$\overline{[x-X(t)]^2} = h\cdot\{\text{Spur}\,(YY^+) + M(t,h)\},$$

where $M(t,h)=O(h^{\frac{1}{2}})$ is uniform on t and $M(t,h)=0$ for $N=0$, i.e., the new functions $\overset{\circ}{\Psi}_{N+1}(x,t)$ satisfy the localization condition (7).

The principal term of the asymptotic representation of the solution to problem (31) has the form

$$\overset{\circ}{\Psi}_1(x,t)=\frac{1}{(2\pi h)^{\frac{m}{4}}}\frac{1}{\sqrt{\det Y}}\exp\left(\frac{i}{h}\right)[S_0(t)+\sum_{\kappa}\dot{X}_\kappa z_\kappa]\cdot\exp\left(\frac{i}{2h}\right)\sum_{\kappa,\ell}\Gamma_{\kappa\ell}(t)(x_\kappa-X_\kappa(t))(x_\ell-X_\ell(t)).$$

In the state $\overset{\circ}{\Psi}_1(x,t)$ the mean values of the particle coordinate

$$\bar{x}_\kappa=\int x_\kappa|\overset{\circ}{\Psi}_1|^2dx=X_\kappa(t)$$

and the particle momentum

$$\bar{p}_\kappa=\int\overset{\circ}{\Psi}{}_1^*\left(-ih\frac{\partial}{\partial x_\kappa}\right)\overset{\circ}{\Psi}_1\,dx=\dot{X}_\kappa(t)$$

coincide with the corresponding classical values of these dynamical variables.

The calculation of the probability density function

$$|\overset{\circ}{\Psi}_1(x,t)|^2=\frac{1}{(2\pi h)^{\frac{m}{2}}}\frac{1}{\sqrt{\det Y}}e^{-\frac{1}{2h}(YY^{+^{-1}}(\vec{x}-\vec{X}),(\vec{x}-\vec{X}))}$$

shows that

$$|\overset{\circ}{\Psi}_1(x,t)|^2\xrightarrow[h\to 0]{}\delta(x-X(t)),$$

i.e., that in the limit the particle exhibits classical behavior.

We have constructed a series that formally satisfies Eq. (14), where the first (principal) term of the series is equal to

$$u_0=\frac{1}{\sqrt{\det Y}}e^{\frac{i}{2}(Y'Y^{-1}\xi,\xi)}$$

In perfectly analogous fashion series are constructed representing formal solutions of Eq. (14) whose first terms are equal to

$$\Lambda_1^{*q_1}\Lambda_2^{*q_2}\cdots\Lambda_m^{*q_m}u_0$$

We shall omit the attendant computations.

LITERATURE CITED

1. Glazman, I. M., Direct Methods of Qualitative Spectral Analysis of Singular Differential Operators, Fizmatgiz, Moscow (1963).
2. Maslov, V. P., On the transition from quantum to classical mechanics in the multidimensional case, Usp. Mat. Nauk, 15(1):91 (1960).
3. Maslov, V. P., Quasi-classical asymptotic representation of certain problems of mathematical physics, Zh. Vychislit. Mat. i Mat. Fiz., Vol. 1, No. 1 (1961).
4. Babich, V. M., Eigenfunctions concentrated in a neighborhood of a closed geodesic, Seminars in Mathematics, Vol. 9: Mathematical Problems in Wave Propagation Theory, Consultants Bureau, New York (1970), pp. 7-26.

FOURIER SERIES AND INTEGRALS ASSOCIATED WITH DUAL EQUATIONS

B. P. Belinskii

We propose to investigate problems involving the determination of a pair of functions one of which belongs to a subspace of a certain Hilbert space, the other to the orthogonal complement, according to a linear relation between the two. We solve some of the problems explicitly and reduce other problems "close" to the latter to equations of the Fredholm type.

In the space $L_2(a, b)$ we introduce two sets of functions:

$$\varepsilon^+ \equiv \{f(x) \mid f(x) = 0, \quad x \in W^-\},$$
$$\varepsilon^- \equiv \{f(x) \mid f(x) = 0, \quad x \in W^+\}, \tag{I}$$

where $(a, b) = W^+ \cup W^-$ and $|a|, |b| < \infty$.

In the article we consider the problem of determining a pair of functions $f^+(x) \in \varepsilon^+$ and $f^-(x) \in \varepsilon^-$ according to the relation

$$Af^+ = f^- + g, \tag{II}$$

where A is a linear operator in $L_2(a, b)$, and $g(x) \in L_2(a, b)$ is assumed to be known.

Let $\{e_n(x)\}_{n=-\infty}^{n=\infty}$ be a complete orthonormalized system in $L_2(a, b)$. If we transform from the functions to their Fourier coefficients according to the system $\{e_n(x)\}$ Eq. (II) is then replaced by the equivalent system

$$(Af^+)_n = f_n^- + g_n \quad (n = -\infty, \ldots, \infty). \tag{II'}$$

For certain special operators A and $e_n(x) = \frac{1}{\sqrt{2\pi}} e^{inx}$, where $(a, b) = (-\pi, \pi)$, we formulate explicit solutions. For operators sufficiently close to A we reduce Eq. (II) [or (II')] to an equivalent equation of the Fredholm type (§ 1). In § 2 we indicate a technique for reducing (II') to an equivalent infinite system of linear algebraic equations. In the same section we investigate a generalization of problem (II). The propositions of § 2 and some of those in § 1 are valid for any complete orthonormalized system $\{e_n(x)\}$.

In § 3 we consider the case $e_n(x) = \frac{1}{\sqrt{2\pi}} e^{inx}$ further for certain operators A. Dual summation equations afford a particular application (§ 4) (see [1, 2] with regard to dual summation equations). In the same section we also analyze a generalization of Eq. (II) to the case $(a, b) = (-\infty, \infty)$.

§ 1

a. Let us consider the ultimately simple case $A = E$. The following is self-evident:

Lemma 1. The sets $\mathcal{E}^+$ and $\mathcal{E}^-$ are subspaces of $L_2(a,b)$. The projectors in $\mathcal{E}^{\pm}$ are operators of multiplication by the characteristic functions $\chi^{\pm}$ of the sets $W^{\pm}$, and $L_2(a,b) = \mathcal{E}^+ \oplus \mathcal{E}^-$.

This solves problem (II) for $A = E$:

$$f^{\pm}(x) = \pm \chi^{\pm}(x) \cdot g(x) \equiv g^{\pm}(x), \tag{1.1}$$

and clearly the solution is unique.

If $e_n(x) = \frac{1}{\sqrt{2\pi}} e^{inx}$, the solution of problem (II') is written in the form

$$f_n^{\pm} = \pm \sum_{m=-\infty}^{\infty} \chi_{n-m}^{\pm} \cdot g_m \equiv g_n^{\pm}, \tag{1.2}$$

where $\chi_m^{\pm}$ is the m-th Fourier coefficient of the functions $\chi^{\pm}(x)$.

Consider the space ℓ_2 of vectors whose components are the Fourier coefficients of functions from $L_2(a,b)$. Accordingly, we denote by $e^{\pm}$ the set of vectors whose components are the Fourier coefficients of functions from $\mathcal{E}^{\pm}$.

Lemma 2. The sets $e^{\pm}$ are subspaces of ℓ_2; $\ell_2 = e^+ \oplus e^-$.

We are now in a position to state a criterion of membership of an element of ℓ_2 in the subspace $e^+(e^-)$:

Lemma 3. $\{f_n\} \in e^{\pm}$ if and only if there is a function $\mathcal{F}(x) \in L_2(a,b)$ such that $f_n = \int_{W^+} \mathcal{F}(x)\, \overline{e_n(x)}\, dx$ for all n.

The necessity is obvious; sufficiency ensues from the equation

$$\sum_{n=-\infty}^{\infty} \int_a^b \varphi(y) \overline{e_n(y)}\, dy \cdot e_n(x) = \varphi(x), \tag{1.3}$$

which is valid for any $\varphi \in L_2(a,b)$.

We shall require this lemma later (§ 3).

The case $A = \alpha E$ $(\alpha = \text{const} \neq 0)$ is treated analogously.

b. Now let $A = E + \varepsilon$ where $\varepsilon = \{\varepsilon_{mn}\}_{m,n=-\infty}^{m,n=\infty}$ is a "small" operator (the precise meaning of the term will be established below); let $e_n(x) = \frac{1}{\sqrt{2\pi}} e^{inx}$. Equation (II') is rewritten in the form

$$f_n^+ = f_n^- + g_n - \sum_{m=-\infty}^{\infty} \varepsilon_{nm} \cdot f_m^+, \tag{1.4}$$

whence, according to (1.2),

$$f_n^+ = g_n^+ - \sum_{K=-\infty}^{\infty} \chi_{n-K}^+ \cdot \sum_{m=-\infty}^{\infty} \varepsilon_{Km} \cdot f_m^+,$$

$$f_n^- = g_n^- + \sum_{K=-\infty}^{\infty} \chi_{n-K}^- \sum_{m=-\infty}^{\infty} \varepsilon_{Km} \cdot f_m^+. \tag{1.5}$$

Note that it suffices to solve the first of Eqs. (1.5), whereupon f_n^- is found from the second equation.

Let all the ε_{km} be so small that

$$c_1 \equiv \sum_{n,m=-\infty}^{\infty} \left| \sum_{k=-\infty}^{\infty} \chi_{n-k}^{+} \varepsilon_{km} \right|^2 < \infty, \tag{1.6}$$

so that the first of Eqs. (1.5) is an equation of the Fredholm type, because condition (1.6) implies complete continuity of the operator

$$\left\{ \sum_{k=-\infty}^{\infty} \chi_{n-k}^{+} \cdot \varepsilon_{km} \right\}_{n,m=-\infty}^{n,m=\infty}$$

(operator of the Hilbert-Schmidt type).

If $c_1 < 1$, the first of Eqs. (1.5) can be solved by successive approximations, and the solution is unique.

This assertion is a special case of the fact that under the condition

$$c_2 \equiv \| \chi^{+}(x) \cdot \varepsilon \| < 1 \tag{1.7}$$

Eq. (II) can be solved by the method of successive approximations [i.e., the specific form of $e_n(x)$ is inessential].

c. Let $e_n(x) = \frac{1}{\sqrt{2\pi}} e^{inx}$ and $A = \{a_{m-n}\}_{m,n=-\infty}^{m,n=\infty}$. Let us assume that $a(x) \neq 0$, $x \in W^+$, where

$$a(x) \equiv \sum_{n=-\infty}^{\infty} a_n \cdot e^{inx}. \tag{1.8}$$

Equation (II') is rewritten as

$$\sum_{m=-\infty}^{\infty} a_{n-m} \cdot f_m^{+} = f_n^{-} + g_n. \tag{1.9}$$

We multiply (1.9) by e^{inx} and sum over all n:

$$a(x)\, f^{+}(x) = f^{-}(x) + g(x), \tag{1.10}$$

whence

$$\begin{aligned} f^{+}(x) &= \frac{g(x)}{a(x)}, \quad x \in W^{+}, \\ f^{-}(x) &= -g(x), \quad x \in W^{-}, \end{aligned} \tag{1.11}$$

so that finally

$$f_n^{+} = \frac{1}{2\pi} \int_{W^+} \frac{g(y)}{a(y)} e^{-iny}\, dy, \qquad f_n^{-} = -\frac{1}{2\pi} \int_{W^-} g(y) e^{-iny}\, dy. \tag{1.12}$$

$$(n = -\infty, \ldots, \infty)$$

It is apparent from (1.12) that we can ease the restriction on $a(x)$, requiring merely that the zeros of $a(x)$ and $g(x)$ coincide (with due regard for multiplicity).

d. The case of an operator A "close" to the one just considered is investigated in a manner analogous to b above. We denote solving operator of problem (1.9) by $\{R^{\pm}_{nm}\}^{n,m=\infty}_{n,m=-\infty}$, so that $f^{\pm}_n = \sum_{m=-\infty}^{\infty} R^{\pm}_{nm} \cdot g_m$. It is easily verified that all the assertions of subsection b are valid, provided the following is assumed there:

$$C_1 \equiv \sum_{n,m=-\infty}^{\infty} \left| \sum_{\kappa} R^{+}_{n\kappa} \cdot \varepsilon_{\kappa m} \right|^2, \tag{1.13}$$

$$C_2 \equiv \| \chi^{+}(x) \cdot \varepsilon \|. \tag{1.14}$$

We assume as before that $a(x) \neq 0,\ x \in W^+$. Note that for $a_{m-n} = \delta_{mn}$ the equations considered in c and d go over to the equations of a and b.

e. All of the foregoing is readily extended to the following cases: the determination of several pairs of functions $f^{\pm}(x)$ [case of a vector space $L_2(a, b)$]; the investigation of more than two sets: $(a, b) = \bigcup_{\kappa=1}^{N} W_\kappa$ and, hence, more than two subspaces ε_κ:

$$\varepsilon_\kappa \equiv \{ f(x) \mid f(x) = 0, \quad x \bar{\in} W_\kappa \}; \tag{1.15}$$

the investigation of a multidimensional space [i.e., (a, b) as a multidimensional cube] and, hence, multidimensional Fourier series.

§ 2

a. We intend to show that Eq. (II) reduces to an operator equation in one unknown function. In this connection we generalize the statement of the problem. Let an arbitrary Hilbert space $\mathcal{H}$ be decomposed into the direct sum of two subspaces $\mathcal{H}^+$ and $\mathcal{H}^-$: $\mathcal{H} = \mathcal{H}^+ \oplus \mathcal{H}^-$. Our problem is to find pairs of elements $f^+ \in \mathcal{H}^+$ and $f^- \in \mathcal{H}^-$ according to the relation

$$A f^+ = f^- + g, \tag{2.1}$$

where A is a linear operator in $\mathcal{H}$ and g is a given element of $\mathcal{H}$.

We introduce the element $f = f^+ - f^-$, i.e., $f^{\pm} = \pm P^{\pm} f$, where $P^{\pm}$ are the projectors in $\mathcal{H}^{\pm}$. Then Eq. (2.1) is transformed to the following:

$$(A P^+ + P^-) f = g. \tag{2.2}$$

For (2.2) we easily deduce the results of §1, a and b.

If $A = E + \varepsilon$ and εP^+ is a completely continuous operator, Eq. (2.2) is an equation of the Fredholm type.

b. Conversely, if for an operator B in $\mathcal{H}$ there exists an operator A in $\mathcal{H}$ such that

$$A P^+ + P^- = B, \tag{2.3}$$

then the infinite system of linear algebraic equations $Bf = g$ reduces to problem (2.1).

c. We observe that the formulations of subsections a and b can be extended to the case of Banach spaces. For example, let S_* be a space of generalized functions over a basic space S. Also, let $S = S^+ \dot{+} S^-$ and

$$S_*^{\pm} \equiv \{f \mid (f, \varphi) = 0, \ \varphi \in S^{\mp}\}. \tag{2.4}$$

It is required to determine a pair of generalized functions f^+ and f^-, belonging to S_*^+ and according to the relation

$$Af^+ = f^- + g. \tag{2.5}$$

§ 3

a. Let $e_n(x) = \frac{1}{\sqrt{2\pi}} e^{inx}$ and $A = \{A_{nm}\}_{n,m=-\infty}^{n,m=\infty}$, where

$$A_{nm} = \begin{cases} \delta_{nm}, & n \geq n_0, \\ -\delta_{nm}, & n < n_0. \end{cases} \tag{3.1}$$

Equation (II') is transformed to the following:

$$\left.\begin{aligned} f_n^+ &= f_n^- + g_n, \quad n \geq n_0, \\ -f_n^+ &= f_n^- + g_n, \quad n < n_0. \end{aligned}\right\} \tag{3.2}$$

Multiplication of Eq. (II') by $e^{-in_0 x}$ makes it possible to have $n_0 = 0$ in (3.2). Let us assume that this transformation has indeed been executed.

We shall show that Eqs. (3.2) can be reduced to the Riemann boundary-value problem with a discontinuous coefficient.

We introduce the following functions of the complex variable z:

$$\begin{aligned} f_>^{\pm}(z) &\equiv \sum_{n \geq 0} f_n^{\pm} z^n, \\ f_<^{\pm}(z) &\equiv \sum_{n < 0} f_n^{\pm} z^n. \end{aligned} \tag{3.3}$$

It is readily verified that the functions $f_>^{\pm}(z)$ are analytic in the disk $|z| < 1$ and that the functions $f_<^{\pm}(z)$ are analytic outside it, where

$$\lim_{z \to \infty} f_<^{\pm}(z) = 0.$$

We assume that

$$\begin{aligned} g_>(z) &\equiv \sum_{n \geq 0} g_m \cdot z^n, \\ g_<(z) &\equiv \sum_{n < 0} g_n \cdot z^n. \end{aligned} \tag{3.4}$$

It is easily verified that $\lim_{z \to \infty} g_<(z) = 0$. From Eqs. (3.2) and the definition of the sets $\varepsilon^{\pm}$ we have

$$
\begin{aligned}
&f_>^+(x)-f_>^-(x)=g_>(x), \quad x\in W^+\cup W^-,\\
&f_<^+(x)+f_<^-(x)=-g_<(x), \quad x\in W^+\cup W^-,\\
&f_>^+(x)=-f_<^+(x) \qquad x\in W^+,\\
&f_>(x)=-f_<^-(x) \qquad x\in W^-.
\end{aligned} \tag{3.5}
$$

From (3.5) we quickly deduce

$$
\left.\begin{aligned}
&f_<^+(x)=f_>^+(x)-g(x), \quad x\in W^-,\\
&f_<^+(x)=-f_>^+(x), \qquad x\in W^+;
\end{aligned}\right\} \tag{3.6}
$$

$$
\left.\begin{aligned}
&f_<^-(x)=f_>^-(x)+g_>(x)-g_<(x), \quad x\in W^+,\\
&f_<^-(x)=-f_>^+(x), \quad x\in W^-.
\end{aligned}\right\} \tag{3.7}
$$

Problems (3.6) and (3.7) constitute the Riemann problem:

Find a pair of functions $f_>^\pm(z)$ and $f_<^\pm(z)$, analytic inside and outside the disk $|z|<1$, respectively, according to relation (3.6) or (3.7), which hold on the boundary of the disk $z=e^{ix}$, $x\in(-\pi,\pi)$. Problems of this type have been investigated in detail in [3]. In particular, the restrictions on the behavior of $f_>^\pm$ and $f_<^\pm$ in the neighborhood of the points of discontinuity of the coefficients of the Riemann problem prove essential. For instance, the requirement of boundedness of the solutions imposes certain conditions on $g_>(x)$ and $g_<(x)$.

Inasmuch as Eqs. (3.6) and (3.7) are not equivalent to Eqs. (3.5), it is required, on solution of the Riemann problem, to insert the result into (3.5).

b. The following case is treated analogously:

$$
A_{nm}=\begin{cases} \delta_{nm}, & n>n_0,\\ -\delta_{nm}, & n<n_0,\\ 0, & n=n_0. \end{cases} \tag{3.8}
$$

Instead of (3.3) we have

$$
f_>^\pm(z)\equiv\sum_{n>0} f_n^\pm\cdot z^n, \quad f_<^\pm(z)\equiv\sum_{n<0} f_n^\pm\cdot z^n. \tag{3.9}
$$

The numbers $f_0^\pm$ enter into the right-hand sides of Eqs. (3.5), (3.6), and (3.7). Let the solution of problem (II') in the case (3.8) be written in the form

$$
f_n^\pm=F_n^\pm(f_0^+, f_0^-, g(x)). \tag{3.10}
$$

Then for the determination of $f_0^\pm$ we have the numerical system of equations

$$
f_0^+=F_0^+(f_0^+, f_0^-, g(x)), \quad f_0^-=F_0^-(f_0^+, f_0^-, g(x)). \tag{3.11}
$$

The following case is treated analogously:

$$A_{nm} = \begin{cases} \delta_{nm}, & n > n_0^1, \\ -\delta_{nm}, & n < n_0^2, \\ 0, & n_0^1 \geq n \geq n_0^2, \end{cases} \tag{3.12}$$

$$|n_0^1|, |n_0^2| < \infty.$$

c. Let us consider $A = \{A_{nm}\}_{n,m=-\infty}^{n,m=\infty}$, where

$$A_{nm} = \delta_{nm} \cdot n^{\varkappa}, \tag{3.13}$$

and $\varkappa > 0$ is an integer. We also assume that $W^+ = \bigcup_{K=1}^{N} (\alpha_K, \beta_K)$. Eq. (II') is rewritten in the form

$$n^{\varkappa} \cdot f_n^+ = f_n^- + g_n \quad (n = -\infty, \ldots, \infty). \tag{3.14}$$

Inasmuch as $\{f_n^{\pm}\}$ and $\{g_n\}$ are elements of ℓ_2, it follows that

$$\sum_{n=-\infty}^{\infty} (1 + n^{2\varkappa}) \cdot |f_n^+|^2 < \infty, \tag{3.15}$$

and, consequently: $f^+(x) \in W_2^{\varkappa}$ [the space of functions that together with their generalized derivatives to and including order $\varkappa$ belong to $L_2(-\pi, \pi)$].

By the embedding theorem of Sobolev [4] $f^+(x) \in C^{(\varkappa-1)}[-\pi, \pi]$ (the space of functions continuous on $[-\pi, \pi]$ with derivatives to and including order $\varkappa - 1$. Consequently, the following relations hold:

$$f^{+(K)}(x) = 0, \quad x \in \overline{W^-}, \quad (0 \leq K \leq \varkappa - 1). \tag{3.16}$$

We introduce the new unknown function

$$\varphi^+(x) \equiv \sum_{n=-\infty}^{\infty} n^{\varkappa} \cdot f_n^+ \cdot e^{inx}, \tag{3.17}$$

whereupon

$$\varphi_n^+ = n^{\varkappa} \cdot f_n^+ \quad (n = -\infty, \ldots, \infty), \tag{3.18}$$

where it is readily verified that $\varphi^+(x) \in \mathcal{E}^+$.

With regard for (3.13) Eq. (3.14) becomes transformed to the following:

$$\varphi_n^+ = f_n^- + g_n \quad (n = -\infty, \ldots, \infty), \tag{3.19}$$

whereupon

$$\varphi_n^+ = g_n^+, \qquad f_n^- = + g_n^- \tag{3.20}$$

We finally have

$$f_n^+ = \frac{g_n^+}{n^{\varkappa}} \quad (n \neq 0), \tag{3.21}$$

where f_0^+ is arbitrary.

We now establish the conditions that must be imposed on $\{g_n^+\}$ in order to obtain the element by Eq. (3.21).

Equations (3.18) and (3.20) lead to the clearly necessary condition

$$g_0^+ = 0, \quad \text{or} \quad \int_{W^+} g^+(t)\,dt = 0. \tag{3.22}$$

Let us introduce the function

$$\mathcal{F}(x) = \begin{cases} 0, & x \in W^-, \\ \int_{\alpha_1}^{x} dx_1 \int_{\alpha_1}^{x_1} dx_2 \cdots \int_{\alpha_1}^{x_{\varkappa-1}} dx_\varkappa\, g^+(x_\varkappa), & x \in W^+. \end{cases} \tag{3.23}$$

It is perceived at once that

$$\mathcal{F}(\alpha_1) = \mathcal{F}'(\alpha_1) = \cdots = \mathcal{F}^{(\varkappa-1)}(\alpha_1) = 0. \tag{3.24}$$

We require that for all $1 \leq \kappa \leq N$ and $1 \leq s \leq \varkappa$

$$\mathcal{J}^s_{\alpha_\kappa, \beta_\kappa} \equiv \int_{\alpha_1}^{\alpha_\kappa, \beta_\kappa} dx_s \int_{\alpha_1}^{x_s} dx_{s+1} \cdots \int_{\alpha_1}^{x_{\varkappa-1}} dx_\varkappa \cdot g^+(x_\varkappa) = 0. \tag{3.25}$$

Then for the same κ and s

$$\mathcal{F}(\alpha_\kappa) = \mathcal{F}'(\alpha_\kappa) = \cdots = \mathcal{F}^{(\varkappa-1)}(\alpha_\kappa) = \mathcal{F}(\beta_\kappa) = \mathcal{F}'(\beta_\kappa) = \cdots = \mathcal{F}^{(\varkappa-1)}(\beta_\kappa) = 0. \tag{3.26}$$

It is readily seen that condition (3.22) follows from conditions (3.26).

Note that in this case $\mathcal{F}(x) \in C^{(\varkappa-1)}(\overline{W^+})$.

Now for $x \in W^+$ we have $\mathcal{F}^{(\varkappa)}(x) = g^+(x)$.

Then from (3.21) and (4.26) we have for $n \neq 0$:

$$\begin{aligned} f_n^+ &= \frac{1}{2\pi \cdot n^\varkappa} \int_{W^+} g^+(x) e^{-inx}\,dx = \frac{1}{2\pi \cdot n^\varkappa} \int_{W^+} \mathcal{F}^{(\varkappa)}(x) e^{-inx}\,dx = \frac{1}{2\pi n^\varkappa}\Big[\Big|_{\partial W^+} \mathcal{F}^{(\varkappa-1)}(x) \cdot e^{-inx} + in\mathcal{F}^{(\varkappa-2)}(x) \cdot e^{-inx} + \\ &+ \cdots + (in)^{\varkappa-1} \cdot \mathcal{F}'(x) e^{-inx} + \int_{W^+} (in)^\varkappa \cdot \mathcal{F}(x) e^{-inx}\,dx \Big] = \frac{i^\varkappa}{2\pi} \int_{W^+} \mathcal{F}(x) e^{-inx}\,dx. \end{aligned} \tag{3.27}$$

It suffices to put (see Lemma 3 of §1)

$$f_0^+ = \frac{i^\varkappa}{2\pi} \int_{W^+} \mathcal{F}(x)\,dx; \tag{3.28}$$

then $f^+(x) = i^\varkappa \cdot \mathcal{F}(x) \in \mathcal{E}^+$, i.e., conditions (3.25) are sufficient.

We now prove their necessity. Let problem (3.14) be solvable; then relations (3.16) are applicable. Let us consider any of the integrals (3.25). We have

$$\mathcal{J}^s_{\alpha_\kappa} = \int_{\alpha_1}^{\alpha_\kappa} dx_s \cdots \int_{\alpha_1}^{x_{\varkappa-1}} dx_\varkappa \cdot g^+(x_\varkappa) = \int_{\alpha_1}^{\alpha_\kappa} dx_s \cdots \int_{\alpha_1}^{x_{\varkappa-1}} dx_\varkappa \sum_{n=-\infty}^{\infty} n^\varkappa \cdot f_n^+ \cdot e^{inx_\varkappa} \equiv \sum_{n=-\infty}^{\infty} n^\varkappa \cdot f_n^+ \cdot \frac{\Phi_s(e^{in\alpha_\kappa}, e^{in\alpha_1})}{n^{\varkappa-s+1}}, \tag{3.29}$$

where Φ_s is a polynomial of order s with coefficients that are independent of n; here $n=0$ is a zero of Φ_s of order $\varkappa-s+1$, so that $\frac{\Phi_s(e^{in\alpha_\kappa}, e^{in\alpha_1})}{n^{\varkappa-s+1}}$ for $n=0$ must be interpreted as the limit. We set

$$\frac{\Phi_s(e^{in\alpha_\kappa}, e^{in\alpha_1})}{n^{\varkappa-s+1}} \equiv \sum_{j=1}^{\varkappa-s+1} \frac{A_j e^{in\gamma}}{n^j}, \tag{3.30}$$

where γ assumes one of the values α_κ, α_1.

Then:

$$\mathcal{J}^s_{\alpha_\kappa} = \sum_{j=1}^{\varkappa-s+1} \frac{A_j \cdot e^{in\gamma}}{n^j} \sum_{n=-\infty}^{\infty} n^{\varkappa} \cdot f_n^+ = \sum_{j=1}^{\varkappa-s+1} A_j \int_{W^+} f^+(x)\delta^{(\varkappa-j)}(x-\gamma)\,dx, \tag{3.31}$$

where it is recognized that for $x \in (-\pi, \pi)$

$$\sum_{n=-\infty}^{\infty} n^{\varkappa-j} e^{in(\gamma-x)} = 2\pi \cdot \delta^{(\varkappa-j)}(x-\gamma). \tag{3.32}$$

From conditions (3.16) we obtain

$$\mathcal{J}^s_{\alpha_\kappa} = 0. \tag{3.33}$$

d. If $g(x) \in W_2^{\varkappa}$ where $\varkappa > 0$ is an integer, then the investigation of problem (II') with

$$A_{nm} = \delta_{nm} \cdot n^{-\varkappa}$$

is analogous to the preceding.

e. Operators close to the A_{nm} of c and d can be treated in a manner analogous to the investigation in §1.

§ 4

a. A number of problems in mathematical physics with mixed boundary and initial conditions produce dual summation equations; it is required to find numbers X_n, $-\infty < n < \infty$, if $\{X_n\} \in \ell_2$ and

$$\sum_{n=-\infty}^{\infty} X_n e_n(x) = \mathcal{F}(x), \quad x \in W^-,$$
$$\sum_{n=-\infty}^{\infty} (AX)_n \cdot e_n(x) = G(x), \quad x \in W^+, \tag{4.1}$$

where A is an operator in ℓ_2; $\{e_n(x)\}_{n=-\infty}^{n=\infty}$ is a complete orthonormalized system of functions in $L_2(a,b)$, $\mathcal{F}(x)$ and $G(x)$ are given elements of $L_2(a,b)$, and $(a,b) = W^+ \cup W^-$.

In terms of the subspaces $e^{\pm}$ Eqs. (4.1) are rewritten in the form

$$X_n = \mathcal{F}_n + f_n^+,$$
$$(AX)_n = G_n + f_n^-, \quad (n = -\infty, \dots, \infty), \tag{4.2}$$

where $\{f_n^{\pm}\} \in e^{\pm}$.

Eliminating X_n from relations (4.2), we arrive at an equation of the type (II'):

$$(Af^+)_n = f_n^- + G_n - A\mathcal{F}_n \quad (n = -\infty, \dots, \infty). \tag{4.3}$$

Thus we have solved the dual summation equations for all cases in which problem (II') is solved. In particular, the system of equations (4.2) can be reduced to an infinite system of linear algebraic equations.

If we consider $e_n(x) = \frac{1}{\sqrt{2\pi}} e^{inx}$ and impose the condition that $f^{\pm}(x)$ be even (odd), we obtain equations of the type (II'), but with $\{\cos nx\}(\{\sin nx\})$ as the complete system. Equations (II') can be treated analogously for $e_n(x) = \frac{1}{\sqrt{2\pi}} \cos(n + \frac{1}{2})x$ or $e_n(x) = \frac{1}{\sqrt{2\pi}} \sin(n + \frac{1}{2})x$. These equations can be solved by analogy with § 3 for $A_{nm} = \delta_{nm} n^{\varkappa}$, where $\varkappa$ is an integer.

For $\varkappa = \pm 1$ problems of this nature have been analyzed in [1, 2].

b. Let Φ be a unitary operator in $L_2(E^n)$. Let us introduce in $L_2(E^n)$ two sets of functions:

$$\varepsilon^{\pm} \equiv \{f(\lambda) \mid (\Phi f)(x) = 0, \ x \in W^{\mp}\}, \tag{4.4}$$

where $E^n = W^+ \cup W^-$.

It is easily verified that the $\varepsilon^{\pm}$ are subspaces of $L_2(E^n)$, that

$$L_2(E^n) = \varepsilon^+ \oplus \varepsilon^- \tag{4.5}$$

and that $f(\lambda) \in \varepsilon^{\pm}$ if and only if

$$f(\lambda) = \Phi^* \chi^{\pm} \Psi(x), \tag{4.6}$$

where $\Psi(x) \in L_2(E^n)$ and $\chi^{\pm}$ is the operator of multiplication by the characteristic function of $W^{\pm}$ (analog of Lemma 3 in § 1).

The problem of determining the pair of functions $f^{\pm}(\lambda) \in \varepsilon^{\pm}$ from the relation

$$Af^+ = f^- + g, \tag{4.7}$$

where A is an operator in $L_2(E^n)$, and g is a given element of $L_2(E^n)$, is thus a special case of the problem considered in § 2, a).

It is well known that if Φ is the Fourier operator and W^+ is the semiaxis $x > 0$, then $f^+(\lambda)$ is analytic in the upper half-plane of the function, and $f^-(\lambda)$ is analytic in the lower half-plane.

Consequently, problem (4.7) is a generalization of the Riemann boundary-value problem for the half-plane [3].

By analogy with § 2, a), problem (4.7) can be reduced to a single equivalent integral equation.

c. We note that if Φ is a bounded operator in $L_2(E^n)$ for which a bounded inverse exists, then all the constructions of subsection b) are repeated verbatim with the exception of Eqs. (4.5) and (4.6), which now have the form

$$L_2(E^n) = \varepsilon^+ \dotplus \varepsilon^-, \tag{4.8}$$

$$f(\lambda) = \Phi^{-1} \chi^{\pm} \Phi \Psi. \tag{4.9}$$

Proceeding from the latter, we can indicate a technique for reducing a broad class of dual integral equations to an equation with a single kernel. Thus, let it be required to determine the function $\chi(\lambda)$ from the dual integral equation

$$\int_{E^n} X(\lambda)\cdot C(\lambda)\cdot \Phi(\lambda,\mu)\,d\lambda = f_1(\mu),\quad \mu\in W^+;$$
$$\int_{E^n} X(\lambda)\cdot \Phi(\lambda,\mu)\,d\lambda = f_2(\mu),\quad \mu\in W, \tag{4.10}$$

where it is known that for a bounded operator with kernel $\Phi(\lambda,\mu)$ there is a bounded inverse; the functions $C(\lambda)$ and $f_{1,2}(\mu)$ are assumed to be known, where $f_{1,2}(\mu)\in L_2(E^n)$. In terms of the subspaces $\mathcal{E}^{(\pm)}$ Eqs. (4.9) may be rewritten in the form

$$\begin{aligned} X(\lambda)\,C(\lambda) &= \tilde{f}_1(\lambda)+f^-(\lambda),\\ X(\lambda) &= \tilde{f}_2(\lambda)+f^+(\lambda), \end{aligned} \tag{4.11}$$

whence we deduce an equation of the form (4.7):

$$C(\lambda)\cdot f^+(\lambda) = f^-(\lambda)+\tilde{f}_1(\lambda)-C(\lambda)\cdot\tilde{f}_2(\lambda). \tag{4.12}$$

It is assumed here that

$$\tilde{f}_{1,2}(\lambda)\equiv\int_{W^\pm} f_{1,2}(\mu)\,\Phi^{-1}(\lambda,\mu)\,d\mu, \tag{4.13}$$

and in the latter that

$$\int_{E^n}\Phi(\lambda,\mu)\,\Phi^{-1}(\sigma,\mu)\,d\mu=\delta(\lambda-\sigma), \tag{4.14}$$

where $\Phi^{-1}(\sigma,\mu)$ is the kernel of the operator Φ^{-1}.

Equation (4.12) can be reduced to an equation in the entire space with one unknown function. Another approach to the reduction of the dual equation to an equation on the entire interval is contained in [5].

d) Consider the special case of (4.7)

$$G(x)\,f^+(x)=f^-(x)+g(x),\qquad x\in E^1, \tag{4.15}$$

where $G(x)$ is a measurable bounded function and the existence of finite limits is presumed:

$$\lim_{x\to+\infty}G(x)=\lim_{x\to-\infty}G(x)\equiv G(\infty)\neq 0. \tag{4.16}$$

Moreover, let one of the sets $W^\pm$ (say, W^+) be finite, and let

$$\Gamma(x)\equiv\frac{G(x)-G(\infty)}{G(\infty)}\in L_2(E^1). \tag{4.17}$$

As in § 2, a), we reduce problem (4.15) to an equivalent integral equation in one unknown function. We put

$$f(x)=G(\infty)\cdot f^+(x)-f^-(x). \tag{4.18}$$

It is readily verified that the functions $f^\pm(x)$ are determined uniquely from (4.18), viz.:

$$f^+=\frac{1}{G(\infty)}\Phi^*\chi^+\Phi f,\qquad f^-=-\Phi^*\chi^-\Phi f. \tag{4.19}$$

Equation (4.15) with regard for (4.19) is transformed to the following:

$$f(x)+\Gamma(x)\frac{1}{2\pi}\int_{W^+} e^{-i\lambda(x-t)}d\lambda\cdot\int_{-\infty}^{+\infty} f(t)dt=g(x),\ x\in E^1. \tag{4.20}$$

The kernel $K(x,t)$ of the integral operator in (4.20) is of the Hilbert-Schmidt type:

$$\iint_{-\infty}^{+\infty}|K(x,t)|^2dx\,dt=\frac{1}{4\pi^2}\int_{-\infty}^{+\infty}|\Gamma(x)|^2 e^{-i\lambda x+i\tilde{\lambda}x}\,dx\int_{-\infty}^{+\infty}\int_{W^+} e^{i\lambda t}d\lambda\cdot\int_{W^+} e^{-i\tilde{\lambda}t}d\tilde{\lambda}\,dt=$$

$$=\frac{1}{2\pi}\int_{-\infty}^{+\infty}|\Gamma(x)|^2 e^{ix(\tilde{\lambda}-\lambda)}dx\int_{W^+}\int_{W^+}\delta(\lambda-\tilde{\lambda})d\lambda d\tilde{\lambda}=\frac{1}{2\pi}\,\mathrm{mes}\,W^+\cdot\|\Gamma\|^2_{L_2(E^1)}<\infty. \tag{4.21}$$

Thus, (4.20) is an equation with a completely continuous operator. It is seen at once that under fulfillment of the inequality

$$\frac{1}{2\pi}\,\mathrm{mes}\,W^+\cdot\|\Gamma\|^2_{L_2(E^1)}<1. \tag{4.22}$$

Eq. (4.29) can be solved by the method of successive approximations.

LITERATURE CITED

1. Noble, B., Some dual series equations involving Jacobi polynomials, Proc. Cambridge Phil. Soc., 59(2):363-371 (1963).
2. Babloyan, A. A., Solution of some dual series, Dokl. Akad. Nauk Armyansk. SSR, 39(3):149-157 (1964).
3. Gakhov, F. D., Boundary Problems, GIFML, Moscow (1963).
4. Smirnov, V. I., Course in Higher Mathematics, Vol. 5.
5. Tseitlin, A. I., On the method of dual integral equations and dual series and its applications to problems of mechanics, Prikl. Mat. Mekh., 30(2):259-270.

A ONE-DIMENSIONAL INVERSE BOUNDARY-VALUE PROBLEM FOR A SECOND-ORDER HYPERBOLIC EQUATION

A. S. Blagoveshchenskii

In the present note we examine the following problem: Find a solution $u(x^1, \ldots, x^n, z, t) \equiv u(x, z, t)$ to the equation

$$u_{tt} = Lu = \frac{\partial}{\partial z}\left[a_0^2(z)\frac{\partial u}{\partial z}\right] + \frac{\partial}{\partial z}\left[a_1(z, D)u\right] + a_1(z, D)\frac{\partial u}{\partial z} + a_2(z, D)u +$$

$$+ b_0(z)\frac{\partial u}{\partial z} + b_1(z, D)u + c_0(z)u \qquad (t > 0,\ z > 0,\ x \in R_n) \tag{1}$$

satisfying the conditions

$$u_z\big|_{z=0} = 0, \tag{2}$$

$$u\big|_{t=0} = 0, \quad u_t\big|_{t=0} = \delta(z)\delta(x). \tag{3}$$

Here, as throughout the remainder of the discussion, we use symbols with subscripts 0, 1, 2 to denote homogeneous polynomials of D of degree 0, 1, and 2, respectively (where $D = \{\frac{\partial}{\partial x^1}, \ldots, \frac{\partial}{\partial x^n}\}$), the coefficients of which are sufficiently smooth functions of z. The polynomials $a_i(z, D)$ are assumed to be such that the operator L is elliptic. We interpret $\delta(z)$ in condition (3) as the quantity $\lim_{z_0 \to +0} \delta(z - z_0)$.

We refer to (1)-(3) as the direct problem. Together with the direct problem we consider the inverse problem, which is to determine the presumably unknown coefficients of the polynomials (or, as amounts to the same thing, the polynomials themselves a_0, a_1, a_2, b_0, b_1, and c_0 under the condition that either the value of the solution $u(x, z, t)$ to problem (1)-(3) for $z = 0$:

$$u(x, z, t)\big|_{z=0} = f(x, t), \tag{4}$$

or certain properties of the function $f(x, t)$ are given.

We shall prove that the solution to the inverse problem (1)-(4) is nonunique. This violation of uniqueness is a simple consequence of Theorem 1 below, which determines which combinations of coefficients of Eq. (1) should logically be sought in attempting to solve the inverse problems.

Theorem 1. Let $u(x, z, t)$ be a solution of the direct problem (1)-(3), and let $f(x, t)$ be the function defined in terms of $u(x, z, t)$ by Eq. (4). Then for the single-valued determination of $f(x,t)$ it is sufficient to specify the following homogeneous polynomials:

$$q_0(y)=\frac{1}{4}\left(\frac{a_0'(y)}{a_0(y)}\right)^2-\frac{1}{2}\frac{a_0''(y)}{a_0(y)}-\frac{1}{2}\frac{b_0'(y)}{a_0(y)}-\frac{b_0^2(y)}{4a_0^2(y)}+c_0(y), \tag{5}$$

$$q_1(y,\kappa)=-\frac{b_0(y)}{a_0^2(y)}a_1(y,\kappa)+b_1(y,\kappa), \tag{6}$$

$$q_2(y,\kappa)=-\left(\frac{1}{a_0^2(y)}a_1^2(y,\kappa)-a_2(y,\kappa)\right), \tag{7}$$

as well as

$$\lambda_0=-\frac{1}{2}\left(\frac{a_0'(y)}{a_0(y)}+\frac{b_0(y)}{a_0(y)}\right)\Big|_{y=0}, \tag{8}$$

$$\lambda_1(\kappa)=-\frac{1}{a_0(y)}a_1(y,\kappa)\Big|_{y=0}, \tag{9}$$

$$\mu_0=\frac{1}{a_0(y)}\Big|_{y=0}, \tag{10}$$

where the coordinate y is related to z by the relation

$$y(z)=\int_0^z\frac{dz}{a_0(z)} \tag{11}$$

For brevity we have denoted the quantities $a_0(y)$, $a_1(y,\kappa)$, etc. by $a_0(z(y))$, $a_1(z(y),\kappa)$, etc. The primes indicate differentiation with respect to y.

Proof. Let $u(\kappa,y,t)$ be the Fourier transform of $u(x,z(y),t)$ on the variables x. It is readily shown that $u(\kappa,y,t)$ is a solution to the problem

$$u_{tt}=u_{yy}+\frac{a_0'(y)}{a_0(y)}u_y-\frac{i}{a_0(y)}\frac{\partial}{\partial y}\left[a_1(y,\kappa)u\right]-\frac{i}{a_0(y)}a_1(y,\kappa)u_y-$$
$$-a_2(y,\kappa)u+\frac{b_0(y)}{a_0(y)}u_y-ib_1(y,\kappa)+c_0(y)u, \tag{12}$$

$$u_y\big|_{y=0}=0, \tag{13}$$

$$u\big|_{t=0}=0, \qquad u_t\big|_{t=0}=\mu_0\,\delta(y). \tag{14}$$

We introduce the new unknown function $v(\kappa,y,t)$ through the equation

$$u(\kappa,y,t)=\varkappa(\kappa,y)\,v(\kappa,y,t), \tag{15}$$

where $\varkappa(\kappa, y)$ is chosen so as to cause the coefficient of $v_y(\kappa,y,t)$ in the equation for v to vanish. It is seen at once that if we normalize $\varkappa(\kappa,y)$ by the condition $\varkappa(\kappa,0) = 1$, then $\varkappa(\kappa,y)$ is uniquely determined in terms of the coefficients of Eq. (12):

$$\varkappa(\kappa,y)=[a_0(y)\mu_0]^{-\frac{1}{2}}\exp\left[i\int_0^y \frac{a_1(y,\kappa)}{a_0(y)}dy-\frac{1}{2}\int_0^y\frac{b_0(y)}{a_0(y)}dy\right]. \tag{16}$$

Now problem (12)-(14) is transformed to the following:

$$v_{tt} = v_{yy} + q(\kappa,y)v, \tag{17}$$

$$v_y + \lambda(\kappa)v\big|_{y=0} = 0, \tag{18}$$

$$v\big|_{t=0} = 0, \quad v_t\big|_{t=0} = \mu_0\delta(y), \tag{19}$$

where

$$q(\kappa,y) = q_0(y) - iq_1(y,\kappa) - q_2(y,\kappa), \tag{20}$$

$$\lambda(\kappa) = \lambda_0 - i\lambda_1(\kappa). \tag{21}$$

It is obvious that $v(\kappa, y, t)$ is uniquely determined by problem (17)-(19) and, hence, by the quantities (5)-(10). However,

$$v(\kappa,0,t) = u(\kappa,0,t) = f(\kappa,t),$$

where $f(\kappa,t)$ is the Fourier transform of $f(x,t)$. This proves Theorem 1.

Remark. The quantities $q(\kappa,y)$ and $\lambda(\kappa)$ are uniquely determined from the coefficients of Eq. (1); the converse is clearly not true, because for given $q(\kappa,y)$ and $\lambda(\kappa)$ it is possible to make an arbitrary choice of, say, the quantities $a_0(y)>0$, $a_1(y,\kappa)$ and $b_0(y)$ so that $a_0(0)=\frac{1}{\mu_0}$, $a_1(0,\kappa) = -\frac{\lambda_1(\kappa)}{\mu_0}$ whereupon the remaining coefficients of Eq. (1) are uniquely determined from $q(\kappa,y)$ and $\lambda(\kappa)$.

Let $f_0(t)$, $f_1(t,\kappa)$, and $f_2(t,\kappa)$ be the first terms of the decomposition of $f(\kappa,t)$ into a power series on $-i\kappa$:*

$$f(\kappa,t) = f_0(t) - if_1(t,\kappa) - f_2(t,\kappa) + \ldots, \tag{22}$$

where $f_0(t)$ satisfies the Krein condition [2]: The operator $\mu_0 E+H: L_2(0,\infty) \to L_2(0,\infty)$ [where H is an integral operator with kernel $f_0'(t+s) + f_0'(|t-s|)$] is positive definite, and $f_1(t,\kappa)$ and $f_2(t,\kappa)$ are arbitrary homogeneous polynomials on κ, depending sufficiently smoothly on t such that $f_1(0,\kappa)=0$, and $f_2(0,\kappa) = f_2'(0,\kappa) = 0$. Further, let $v_0(y,t)$, $v_1(y,t,\kappa)$, and $v_2(y,t,\kappa)$ be the first terms of the decomposition of $v(\kappa,y,t)$ in a power series on $-i\kappa$:

$$v(\kappa,y,t) = v_0(y,t) - iv_1(y,t,\kappa) - v_2(y,t,\kappa) + \ldots, \tag{23}$$

Substituting (22) and (23) into (17)-(19) and equating terms with like powers of κ, we obtain a recursive sequence of problems for $v_0(y,t)$, $v_1(y,t,\kappa)$, and $v_2(y,t,\kappa)$ for which the following is true:

Theorem 2. *Let there be given $f_0(t)$, $f_1(t,\kappa)$, $f_2(t,\kappa)$ satisfying the above-stated conditions. Then there also exists a unique set of variables (5)-(10) such that the solution of the above sequence of problems satisfies the conditions*

---

* The coefficients of the polynomials $f_0(t)$, $f_1(t,\kappa)$, and $f_2(t,\kappa)$ are the zeroth, first, and second moments of $f(x,t)$ on x, so that the ensuing results differ only in form from those of [1].

$$\left.\begin{array}{l} v_0(y,t)\big|_{y=0} = f_0(t),\; v_1(y,t,\kappa)\big|_{y=0} = f_1(t,\kappa), \\ v_2(y,t,\kappa)\big|_{y=0} = f_2(t,\kappa). \end{array}\right\} \tag{24}$$

Remark. Theorem 2 is not equivalent to the assertion that there is a solution to the inverse problem (1)-(3), (24), because the transition from this problem to the recursive sequence of problems considered in Theorem 2 is correct under the condition that Eq. (1) is hyperbolic or, equivalently, that the quadratic form $q_2(y, \kappa)$ is positive definite. In fact, the existence and finiteness of the solution $u(x, z, t)$ on x and, hence, the applicability of the subsequent arguments can be guaranteed only in the case of the hyperbolic equation (1). The question as to the conditions on $f_0(t)$, $f_1(t, \kappa)$, and $f_2(t, \kappa)$ such that the form $q_2(y, \kappa)$ will be positive definite remains unresolved at this time.

LITERATURE CITED

1. Blagoveshchenskii, A. S. in: Topics in Mathematical Physics, Vol. 1, Consultants Bureau, New York (1967), pp. 55-67.
2. Krein, M. G., Dokl. Akad. Nauk SSSR, 94(6):987-990 (1954).

RAYLEIGH WAVES CONCENTRATED NEAR A RAY ON THE SURFACE OF AN INHOMOGENEOUS ELASTIC BODY

N. Ya. Kirpichnikova

We propose by the parabolic equation method to formulate the high-frequency asymptotic behavior (as the frequency $\omega \to \infty$) of generalized steady-state Rayleigh waves concentrated near a particular ray $\mathcal{L}$ at the Rayleigh velocity. We assume that $\mathcal{L}$ lies entirely on the surface S of an inhomogeneous elastic body Ω.

Rayleigh waves are found as the sum of the longitudinal and transverse harmonic solutions of the dynamical equations of elasticity theory with complex eikonals, where those solutions are represented as series in reciprocal powers of the parameter ω and formally satisfy the stated equations and boundary condition of zero stress on S.

In the formulation of the solution we use certain methods of [1, 2, 3].

§1. The Curvilinear Ray System of Coordinates and Auxiliary Problems

Consider an inhomogeneous elastic body Ω of arbitrary shape bounded by a sufficiently smooth stress-free surface S. For the ray $\mathcal{L}$ on S we adopt some extremal of the Fermat integral

$$\int \frac{ds}{c(x, y, z)}, \quad (x, y, z) \in S, \tag{1}$$

where ds is an element of length of the curve of integration and $c(x, y, z)$ is the Rayleigh velocity, determined from the equation

$$\left(\frac{1}{b^2} - \frac{2}{c^2}\right)^2 - \frac{4}{c^2}\sqrt{\frac{1}{c^2} - \frac{1}{a^2}}\ \sqrt{\frac{1}{c^2} - \frac{1}{b^2}} = 0,$$

in which a and b denote the longitudinal and transverse wave velocities, respectively.

We introduce the curvilinear coordinate system $q^1 = \tau$, $q^2 = \alpha$, $q^3 = \nu$, which is essentially attached to $\mathcal{L}$.

For any point M_1 of the ray $\mathcal{L}$ we determine the coordinate τ from the formula

$$\tau = \int_{M_0}^{M_1} \frac{ds}{c(x, y, z)}, \quad \cup M_0 M_1 \in \mathcal{L} \subset S,$$

where M_0 is a fixed point on $\mathcal{L}$.

From every point M_1 of $\mathcal{L}$ we draw the extremals of integral (I) that intersect $\mathcal{L}$ orthogonally.

We shall characterize any point M_2 of S by the coordinates τ, α, where the parameter α is equal to

$$\alpha = \int_{M_1}^{M_2} \frac{ds}{c(x,y,z)}, \quad \cup M_1 M_2 \in S,$$

the integration extending along the extremal that runs through the point M_2 and intersects $\mathcal{L}$ at M_1. The equation $\alpha = 0$ defines the original ray $\mathcal{L}$.

The parameter ν of any point M of the body Ω is the distance along the normal from M to the corresponding point M_2 of S.

In our curvilinear coordinate system the following formula holds for the radius vector $\vec{x}$ of the point M:

$$\vec{x} = \vec{r}(\tau, \alpha) + \nu \vec{n}(\tau, \alpha),$$

where (τ, α) is a point M_2 on S such that the normal $\vec{n}(\tau, \alpha)$ passes through M, and $\vec{r}(\tau, \alpha)$ is the radius vector of M_2 The points of the elastic body correspond to values of $\nu > 0$. The condition $\nu = 0$ determines the points of S, whose parametric equation has the form $\vec{r} = \vec{r}(\tau, \alpha)$.

The parameters τ, α, and ν form a curvilinear coordinate system $q^1 = \tau$, $q^2 = \alpha$, $q^3 = \nu$ in the vicinity of S. In general, the coordinate system τ, α, ν is not orthogonal.

The dynamical equations of elasticity theory have the following form in an arbitrary curvilinear coordinate system q^1, q^2, q^3:

$$\sigma_{ij} \sqrt{g} \frac{\partial G^{ij}}{\partial q^s} + 2 \frac{\partial}{\partial q^i} \left(\sigma_{sj} G^{ij} \sqrt{g} \right) - 2 \rho G_{is} \sqrt{g} \frac{\partial^2 \varphi^i}{\partial t^2} = 0 ; \quad i, j, s = 1, 2, 3. \tag{2}$$

In Eq. (2) we denote by G_{ij} the coefficients of a metric tensor in the curvilinear coordinates q^1, q^2, q^3, so that the square of the length element ds is equal to

$$ds^2 = (d\vec{x} \cdot d\vec{x}) = G_{ij} dq^i dq^j, \quad i, j = 1, 2, 3.$$

The components of the stress tensor σ_{ij} and the displacement vector $\vec{u} = (\varphi^1, \varphi^2, \varphi^3)$ for the given isotropic elastic medium are related by the expressions

$$\sigma_{ij} = \frac{\lambda}{\sqrt{g}} \frac{\partial}{\partial q^s} \left(\sqrt{g} \varphi^s \right) G_{\kappa j} \delta^\kappa_i + \mu \left[\frac{\partial G_{ij}}{\partial q^s} \varphi^s + G_{is} \frac{\partial \varphi^s}{\partial q^j} + G_{js} \frac{\partial \varphi^s}{\partial q^i} \right]. \tag{3}$$

In Eqs. (2) and (3) $\lambda = \lambda(q^1, q^2, q^3)$ and $\mu = \mu(q^1, q^2, q^3)$ are the Lamé parameters, $\rho = \rho(q^1, q^2, q^3)$ is the density of the elastic medium, t is the time, and

$$g = \det \| G_{ij} \| \neq 0.$$

The coefficients of the metric tensor G^{ij} are related to the coefficients of the tensor G_{ij} by the equations

$$G^{i\kappa} G_{\kappa j} = \delta^i_j,$$

where δ^i_j is the Kronecker delta.

The indices i, j, κ, and s run through the values 1, 2, and 3 independently of one another.

Here and elsewhere every letter index occurring twice in a product is the summation index over all possible values of that index.

The functions λ, μ, and ρ are assumed to be differentiable as many times as needed, i.e., they have continuously-differentiable partial derivatives to and including a certain order. We shall not give this special mention in each particular case, tacitly assuming by writing down the derivatives of a particular order that they indeed exist and are continuous. The latter remark applies to the coefficients G_{ij} of the metric tensor as well, as the surface S is assumed to be smooth.

As a result of the relations $dx^2 = (d\vec{x}', d\vec{x}')$ we obtain for the coefficients of the metric tensor

$$G_{\tau\tau} = (\vec{x}'_\tau, \vec{x}'_\tau) = g_{\tau\tau} - 2\nu b_{\tau\tau} + \nu^2 (\vec{n}'_\tau, \vec{n}'_\tau),$$

$$G_{\tau\alpha} = (\vec{x}'_\tau, \vec{x}'_\alpha) = g_{\tau\alpha} - 2\nu b_{\tau\alpha} + \nu^2 (\vec{n}'_\tau, \vec{n}'_\alpha),$$

$$G_{\alpha\alpha} = (\vec{x}'_\alpha, \vec{x}'_\alpha) = g_{\alpha\alpha} - 2\nu b_{\alpha\alpha} + \nu^2 (\vec{n}'_\alpha, \vec{n}'_\alpha),$$

$$G_{\tau\nu} = 0, \quad G_{\alpha\nu} = 0, \quad G_{\nu\nu} = 1.$$

In view of the fact that the solutions we seek must be concentrated near $\mathcal{L}$, they will differ significantly from zero only for sufficiently small values of the parameters α and ν. With regard for the smallness of α and ν we obtain for the coefficients of the metric tensor

$$G_{\tau\tau} = g_{\tau\tau} + \frac{\partial g_{\tau\tau}}{\partial \alpha}\alpha - 2\nu b_{\tau\tau} + \frac{1}{2}\frac{\partial^2 g_{\tau\tau}}{\partial \alpha^2}\alpha^2 - 2\frac{\partial b_{\tau\tau}}{\partial \alpha}\nu\alpha + (\vec{n}'_\tau \cdot \vec{n}'_\tau)\nu^2 + O(|\alpha|^3, \alpha^2\nu, |\alpha|\nu^2, \nu^3),$$

$$G_{\tau\alpha} = -2 b_{\tau\alpha}\nu - 2\frac{\partial b_{\tau\alpha}}{\partial \alpha}\nu\alpha + (\vec{n}'_\tau, n'_\alpha)\nu^2 + O(|\alpha|^3, \alpha^2\nu, |\alpha|\nu^2, \nu^3),$$

$$G_{\alpha\alpha} = g_{\alpha\alpha} + \frac{\partial g_{\alpha\alpha}}{\partial \alpha}\alpha - 2 b_{\alpha\alpha}\nu + \frac{1}{2}\frac{\partial^2 g_{\tau\tau}}{\partial \alpha^2}\alpha^2 - 2\frac{\partial b_{\alpha\alpha}}{\partial \alpha}\alpha\nu + (\vec{n}'_\alpha \vec{n}'_\alpha)\nu^2 + O(|\alpha|^3, \alpha^2\nu, |\alpha|\nu^2, \nu^3),$$

$$G_{\tau\nu} = 0, \quad G_{\alpha\nu} = 0, \quad G_{\nu\nu} = 1. \tag{4}$$

All the coefficients of different powers of α and ν in relations (4) are calculated on $\mathcal{L}$, i.e., for $\alpha = \nu = 0$.

We note that

$$g_{\tau\tau}\big|_{\mathcal{L}} = c^2(\tau, 0, 0), \quad g_{\alpha\alpha}\big|_{\mathcal{L}} = c^2(\tau, 0, 0),$$

where the Euler equation for $\mathcal{L}$ has the form

$$\frac{\partial}{\partial \alpha}\left(\frac{g_{\tau\tau}}{c^2}\right)\Bigg|_{\alpha=0,\ \nu=0} = 0 \tag{5}$$

We shall rely heavily on the latter condition in the course of our intermediate calculations.

The problem of wave propagation in an elastic medium reduces mathematically to the problem of integration of the elasticity-theory equations (2) and (3). The boundary conditions of zero stress on the boundary S of the body Ω are written as follows:

$$\sigma_{i\nu}\big|_{\nu=0} = 0, \qquad i = \tau, \alpha, \nu, \tag{6}$$

where $\sigma_{i\nu}$ are the components of the stress tensor.

§ 2. Derivation of Recursion Relations for the Components of the Displacement Vector

We are interested in the asymptotic behavior of generalized steady-state Rayleigh waves concentrated near the ray $\mathcal{L}$ on the assumption that the frequency ω of the wave process is sufficiently high.

We represent the solution $\vec{u}$ of the stated problem as the sum of the longitudinal $(\vec{u}_a)$ and transverse $(\vec{u}_b)$ solutions:

$$\vec{u} = \vec{u}_a + \vec{u}_b, \tag{7}$$

$$\vec{u}_a = \sum_{k=0}^{\infty} \frac{\vec{u}_{ka}}{\omega^{k/2}} \exp\left[-i\omega(t-\tau_a)\right], \tag{8}$$

$$\vec{u}_b = \sum_{k=0}^{\infty} \frac{\vec{u}_{kb}}{\omega^{k/2}} \exp\left[-i\omega(t-\tau_b)\right], \tag{9}$$

where $\vec{u}_{ka}$ and $\vec{u}_{kb}$ are unknown vectors characterizing the longitudinal and transverse modes, respectively, and the functions τ_a and τ_b satisfy the eikonal equations

$$G^{ij} \frac{\partial \tau_a}{\partial q^i} \frac{\partial \tau_a}{\partial q^j} = \frac{1}{a^2}, \tag{10}$$

$$G^{ij} \frac{\partial \tau_b}{\partial q^i} \frac{\partial \tau_b}{\partial q^j} = \frac{1}{b^2}. \tag{11}$$

Proceeding from the analogy with Rayleigh waves, we require

$$\tau_a \big|_{\mathcal{L}} = \tau_b \big|_{\mathcal{L}} = \tau. \tag{12}$$

We shall assume that the solution we seek is concentrated near $\mathcal{L}$ in a strip of order $\omega^{-1/2}$ about the coordinate α. The formulas obtained below bear out the validity of the latter assumption.

We introduce the new variable

$$\eta = \alpha\, \omega^{+1/2}. \tag{13}$$

We write the unknown vectors $\vec{u}_{ka}$, and $\vec{u}_{kb}$ as follows:

$$\begin{gathered} \vec{u}_{ki} = \left(\overset{k}{V}{}_i^{\tau}, \overset{k}{V}{}_i^{\alpha}, \overset{k}{V}{}_i^{\nu} \right), \\ \overset{k}{V}{}_i^{j} = \overset{k}{V}{}_i^{j}(\tau, \eta, \nu), \quad j = \tau, \alpha, \nu, \quad i = a, b. \end{gathered} \tag{14}$$

Inasmuch as we are seeking sufficiently smooth solutions concentrated near $\mathcal{L}$, we can use the eikonal equations (10) and (11) and conditions (12) to find all derivatives computed on $\mathcal{L}$ of the functions τ_a and τ_b with respect to the variables α and ν.

Here we shall give only the expressions for the derivatives of the functions τ_a and τ_b to be used below:

$$\left.\begin{array}{l}
\left.\frac{\partial\tau_a}{\partial\alpha}\right|_{\mathcal{L}} = \left.\frac{\partial\tau_b}{\partial\alpha}\right|_{\mathcal{L}} = \left.\frac{\partial^2\tau_a}{\partial\alpha^2}\right|_{\mathcal{L}} = \left.\frac{\partial^2\tau_b}{\partial\alpha^2}\right|_{\mathcal{L}} = 0, \\
\left.\frac{\partial\tau_a}{\partial\nu}\right|_{\mathcal{L}} = i\left.\sqrt{\frac{1}{c^2}-\frac{1}{a^2}}\right|_{\mathcal{L}}, \quad \left.\frac{\partial\tau_b}{\partial\nu}\right|_{\mathcal{L}} = i\left.\sqrt{\frac{1}{c^2}-\frac{1}{b^2}}\right|_{\mathcal{L}}, \\
\left.\frac{\partial}{\partial\alpha}\left(\frac{\partial\tau_a}{\partial\nu}\right)^2\right|_{\mathcal{L}} = \left.\frac{\partial}{\partial\alpha}\left(\frac{1}{a^2}-G^{\tau\tau}\right)\right|_{\mathcal{L}}, \quad \left.\frac{\partial}{\partial\alpha}\left(\frac{\partial\tau_b}{\partial\nu}\right)^2\right|_{\mathcal{L}} = \left.\frac{\partial}{\partial\alpha}\left(\frac{1}{b^2}-G^{\tau\tau}\right)\right|_{\mathcal{L}}, \\
\left.\frac{\partial}{\partial\nu}\left(\frac{\partial\tau_a}{\partial\nu}\right)^2\right|_{\mathcal{L}} = \left.\frac{\partial}{\partial\nu}\left(\frac{1}{a^2}-G^{\tau\tau}\right)\right|_{\mathcal{L}}, \quad \left.\frac{\partial}{\partial\nu}\left(\frac{\partial\tau_b}{\partial\nu}\right)^2\right|_{\mathcal{L}} = \left.\frac{\partial}{\partial\nu}\left(\frac{1}{b^2}-G^{\tau\tau}\right)\right|_{\mathcal{L}}, \\
\left.\frac{\partial^2}{\partial\alpha^2}\left(\frac{\partial\tau_a}{\partial\nu}\right)^2\right|_{\mathcal{L}} = \left.\frac{\partial^2}{\partial\alpha^2}\left(\frac{1}{a^2}-G^{\tau\tau}\right)\right|_{\mathcal{L}}, \quad \left.\frac{\partial^2}{\partial\alpha^2}\left(\frac{\partial\tau_a}{\partial\nu}\right)^2\right|_{\mathcal{L}} = \left.\frac{\partial^2}{\partial\alpha^2}\left(\frac{1}{b^2}-G^{\tau\tau}\right)\right|_{\mathcal{L}}.
\end{array}\right\} \tag{15}$$

We decompose the coefficients of Eqs. (2) in a power series on α and ν in the vicinity of $\mathcal{L}$, taking account of Eqs. (4). We then substitute the solution (7)-(9), (14) into the resulting equations and set the coefficients equal to zero for different powers of the parameter $\omega^{1/2}$, beginning with ω^2.

Next we consider the direct determination of the components of the vectors (14). This will give us the principal terms in the decompositions (14) (for $\kappa = 0$), and for the remaining terms ($\kappa > 0$) we shall give only recursion relations.

Equations (2) ($s = 1, 3$; order ω^2) yield

$$\begin{aligned}
&\left[c^2 - a^2 - b^2c^2\left(\frac{\partial\tau_a}{\partial\nu}\right)^2\right]\overset{\circ}{V}{}_a^{\tau} - (a^2-b^2)\frac{\partial\tau_a}{\partial\nu}\overset{\circ}{V}{}_a^{\nu} = 0, \\
&-(a^2-b^2)\frac{\partial\tau_a}{\partial\nu}\overset{\circ}{V}{}_a^{\tau} + \left[1 - \frac{b^2}{c^2} - a^2\left(\frac{\partial\tau_a}{\partial\nu}\right)^2\right]\overset{\circ}{V}{}_a^{\nu} = 0,
\end{aligned} \tag{16}$$

where all the coefficients of the system are computed on $\mathcal{L}$. It follows from Eqs. (15) that

$$\left.\left(\frac{\partial\tau_a}{\partial\nu}\right)^2\right|_{\mathcal{L}} = \left.\left(\frac{1}{a^2} - \frac{1}{g_{\tau\tau}}\right)\right|_{\mathcal{L}},$$

i.e., the determinant of the system (16), composed of the coefficients of the unknowns $\overset{\circ}{V}{}_a^{\tau}$ and $\overset{\circ}{V}{}_a^{\nu}$, is equal to zero, and the indicated system always has a nontrivial solution. Consequently, from the system (16) we have

$$\overset{\circ}{V}{}_a^{\nu} = c^2\tau_{,\nu}^{a}\,\overset{\circ}{V}{}_a^{\tau}, \tag{17}$$

where

$$\tau_{,\nu}^{a} = \left.\frac{\partial\tau_a}{\partial\nu}\right|_{\mathcal{L}}. \tag{18}$$

From Eqs. (2) ($s = 2$, order ω^2) we obtain

$$\overset{\circ}{V}{}_a^{\alpha} = 0. \tag{19}$$

For the higher-order values of $\overset{\kappa}{V}{}_a^{\tau}$ and $\overset{\kappa}{V}{}_a^{\tau}$, $\kappa > 0$, we find

$$\left. \begin{aligned} a_{11} \overset{\kappa}{V}{}_a^{\tau} + a_{12} \overset{\kappa}{V}{}_a^{\nu} = f_1^{(\kappa)}, \\ a_{21} \overset{\kappa}{V}{}_a^{\tau} + a_{22} \overset{\kappa}{V}{}_a^{\nu} = f_2^{(\kappa)}, \end{aligned} \right\} \tag{20}$$

where

$$\left. \begin{aligned} a_{11} = -\frac{(a^2-c^2)(a^2-b^2)}{a^2}, \quad \tau_{\nu}^{a} = \frac{\partial \tau_a}{\partial \nu}\bigg|_{\mathscr{L}}, \\ a_{21} = a_{12} = -(a^2-b^2)\tau_{\nu}^{a}, \quad a_{22} = \frac{a^2-b^2}{c^2}, \end{aligned} \right\} \tag{21}$$

and $f_1^{(\kappa)}$ and $f_2^{(\kappa)}$ are functions of $\overset{m}{V}{}_a^{\tau}$ and $\overset{m}{V}{}_0^{\nu}$ and their derivatives with respect to τ, η, and ν for $m < \kappa$.

The necessary and sufficient condition for the solvability of the system (20) is written as follows:

$$f_1^{(\kappa)} + c^2 \tau_{\nu}^{a} f_2^{(\kappa)} = 0, \quad \kappa = 1, 2, 3, \ldots . \tag{22}$$

Let $\kappa = 1$. From the system (20) we have

$$\left. \begin{aligned} a_{11} \overset{1}{V}{}_a^{\tau} + a_{12} \overset{1}{V}{}_a^{\nu} = f_1^{(1)}, \\ a_{21} \overset{1}{V}{}_a^{\tau} + a_{22} \overset{1}{V}{}_a^{\nu} = f_2^{(1)}, \end{aligned} \right\} \tag{23}$$

where in Eqs. (23) the coefficients a_{ij} are determined by relations (21) and the functions $f_j^{(1)}$, $j = 1, 2$, have the form

$$f_1^{(1)} = \eta \left\{ -\frac{\partial}{\partial \alpha}\left[g_{\tau\tau} - a^2 - b^2 g_{\tau\tau} (\tau_{\nu}^{a})^2 \right] \overset{0}{V}{}_a^{\tau} + \left(\frac{\partial}{\partial \alpha}\left[(a^2-b^2)\tau_{\nu}^{a} \right] \right) \overset{0}{V}{}_a^{\nu} \right\},$$

$$f_2^{(1)} = \eta \left\{ \frac{\partial}{\partial \alpha}\left[(a^2-b^2)\tau_{\nu}^{a} \right] \overset{0}{V}{}_a^{\tau} - \frac{\partial}{\partial \alpha}\left[1 - b^2 G^{\tau\tau} - a^2 (\tau_{\nu}^{a})^2 \right] \overset{0}{V}{}_a^{\nu} \right\}.$$

Making use of Eqs. (5), (15), and (17), we can easily show that condition (22) holds for $\kappa = 1$, and the system (23) is equivalent to the following relation:

$$\overset{1}{V}{}_a^{\nu} = c^2 \tau_{\nu}^{a} \overset{1}{V}{}_a^{\tau} + \eta \frac{\partial}{\partial \alpha}(c^2 \tau_{\nu}^{a}) \overset{0}{V}{}_a^{\tau}. \tag{24}$$

For the function V_a^{α} from Eq. (2) ($s = 2$, order $\omega^{3/2}$) we have

$$\overset{1}{V}{}_a^{\alpha} = -i \frac{\partial \overset{0}{V}{}_a^{\tau}}{\partial \eta}. \tag{26}$$

Let $\kappa = 2$. The solvability conditions (22) have the form

$$2ic^4 \tau_{,\nu}^{a} \frac{\partial \overset{0}{V}_a^{\tau}}{\partial \nu} + 2ic^2 \frac{\partial \overset{0}{V}_a^{\tau}}{\partial \tau} + c^2 \frac{\partial^2 \overset{0}{V}_a^{\tau}}{\partial \eta^2} + A_a^{\tau} \overset{0}{V}_a^{\tau} = 0, \tag{26}$$

and from the system (20) we obtain

$$\overset{2}{V}_a^{\nu} = c^2 \tau_{,\nu}^{a} \overset{2}{V}_a^{\tau} + \eta \frac{\partial}{\partial \alpha} (c^2 \tau_{,\nu}^{a}) \overset{1}{V}_a^{\tau} + \frac{\eta^2}{2!} \frac{\partial^2}{\partial \alpha^2} (g_{\tau\tau} \tau_{,\nu}^{a}) \overset{0}{V}_a^{\tau} + \frac{\overset{1}{\Lambda}_a (\overset{0}{V}_a^{\tau}, \overset{1}{V}_a^{\tau})}{(a^2 - b^2) \tau_{,\nu}^{a}}. \tag{27}$$

Here $\overset{1}{\Lambda}_a$ is a linear differential operator relative to the functions $\overset{1}{V}_a^{\alpha}$ and $\overset{1}{V}_a^{\tau}$:

$$\overset{1}{\Lambda}_a = a_1 \frac{\partial^2 \overset{0}{V}_a^{\tau}}{\partial \eta^2} + a_2 \frac{\partial \overset{0}{V}_a^{\tau}}{\partial \nu} + a_3 \frac{\partial \overset{0}{V}_a^{\tau}}{\partial \tau} + a_4 \overset{0}{V}_a^{\tau} + a_5 \eta \frac{\partial \overset{1}{V}_a^{\tau}}{\partial \eta} + a_6 \overset{1}{V}_a^{\tau}, \tag{28}$$

where A_a^{τ}, a_1, a_2, a_3, a_4, a_5, and a_6 are functions of τ The component $\overset{\kappa}{V}_a^{i}$ is determined by the formula

$$\overset{2}{V}_a^{\alpha} = (a_7 + a_8 \nu) \overset{0}{V}_a^{\tau}, \quad a_7 = a_7(\tau), \quad a_8 = a_8(\tau). \tag{29}$$

We shall not write out the explicit expressions for the functions A_a^{τ}, a_i in the intermediate calculations due to their complex form.

Analogously, from the system (20) for $\kappa \geq 3$ we can find formulas for the coefficients ($i = \tau, \alpha, \nu, \kappa \geq 3$) of the higher approximation.

Our next task is to formulate recursion relations for the components of the vector $\vec{u}_b$.

For the functions $\overset{0}{V}_b^{\tau}$ and $\overset{0}{V}_b^{\alpha}$ we obtain the linear system

$$\left. \begin{aligned} b_{11} \overset{0}{V}_b^{\tau} + b_{12} \overset{0}{V}_b^{\nu} &= 0, \\ b_{21} \overset{0}{V}_b^{\tau} + b_{22} \overset{0}{V}_b^{\nu} &= 0. \end{aligned} \right\} \tag{30}$$

The coefficients b_{ij} are calculated on $\mathcal{L}$ and are equal to

$$\begin{aligned} b_{11} &= -(a^2 - b^2), \quad b_{22} = -(a^2 - b^2)(\tau_{,\nu}^{b})^2, \\ b_{12} &= b_{21} = -(a^2 - b^2) \tau_{,\nu}^{b}, \end{aligned} \tag{31}$$

while the determinant of the system (30) is equal to zero.

From Eqs. (30) we have

$$\overset{0}{V}_b^{\tau} = -\tau_{,\nu}^{b} \overset{0}{V}_b^{\nu}, \quad \tau_{,\nu}^{b} = \left. \frac{\partial \tau_b}{\partial \nu} \right|_{\mathcal{L}}. \tag{32}$$

For the functions V_b^{α} we obtain

$$\overset{0}{V}_b^{\alpha} \neq 0. \tag{33}$$

In the next higher approximations ($\kappa = 1, 2, \ldots$) we find for $\overset{\kappa}{V}_b^{\tau}$ and $\overset{\kappa}{V}_b^{\nu}$,

$$\left.\begin{aligned} b_{11}\overset{\kappa\,\tau}{V_b} + b_{12}\overset{\kappa\,\nu}{V_b} &= g_1^{(\kappa)}, \\ b_{21}\overset{\kappa\,\tau}{V_b} + b_{22}\overset{\kappa\,\nu}{V_b} &= g_2^{(\kappa)}. \end{aligned}\right\} \tag{34}$$

Here the coefficients b_{ij} are determined by Eqs. (31), and $g_1^{(\kappa)}$ and $g_2^{(\kappa)}$ are functions of $\overset{m\,\tau}{V_b}$ and $\overset{m\,\nu}{V_b}$ and their derivatives with respect to τ, η, and ν for $m < \kappa$.

The necessary and sufficient condition for the solvability of the system (34) has the form

$$g_2^{(\kappa)} - \tau_\nu^b\, g_1^{(\kappa)} = 0, \quad \kappa = 1, 2, \ldots \tag{35}$$

For $\kappa = +1$ condition (35) is automatically fulfilled, and from (34) we obtain

$$\overset{1\,\tau}{V_b} = -\tau_\nu^b \overset{1\,\nu}{V_b} - \eta\left(\frac{\partial}{\partial\alpha}\tau_\nu^b\right)\overset{0\,\nu}{V_b}. \tag{36}$$

After a series of intermediate calculations we have from (34) for $\kappa = 2$

$$\overset{2\,\tau}{V_b} = -\tau_\nu^b \overset{2\,\nu}{V_b} - \eta\left(\frac{\partial}{\partial\alpha}\tau_\nu^b\right)\overset{1\,\nu}{V_b} - \frac{\eta^2}{2!}\left(\frac{\partial^2}{\partial\alpha^2}\tau_\nu^b\right)\overset{0\,\nu}{V_b} + \frac{\Lambda_b^1\left(\overset{0\,\nu}{V_b}, \overset{1\,\nu}{V_b}, \overset{0\,\alpha}{V_b}, \overset{1\,\alpha}{V_b}\right)}{a^2 - b^2}, \tag{37}$$

where, analogously to (28), Λ_b^1 is a linear differential operator relative to the functions $\overset{0\,\nu}{V_b}$, $\overset{1\,\nu}{V_b}$ and $\overset{0\,\alpha}{V_b}$, $\overset{1\,\alpha}{V_b}$.

Condition (35) for the solvability of the system (34) for $\kappa = 2$ is rewritten in the form

$$2i\,\frac{\partial \overset{0\,\nu}{V_b}}{\partial\tau} + \frac{\partial^2}{\partial\eta^2}\overset{0\,\nu}{V_b} + 2ic^2\tau_\nu^b\,\frac{\partial \overset{0\,\nu}{V_b}}{\partial\nu} + A_b^\nu\,\overset{0\,\nu}{V_b} = 0, \qquad A_b^\nu = A_b^\nu(\tau). \tag{38}$$

We find the functions $\overset{0\,\alpha}{V_b}$ from the equation

$$\Lambda_b^2\left(\overset{0\,\alpha}{V_b}\right) + a_9\,\overset{0\,\nu}{V_b} = 0, \quad a_9 = a_9(\tau). \tag{39}$$

In Eq. (39) Λ_b^2 is a linear differential operator such that on the surface S it is equal to

$$\Lambda_b^2\Big|_S = a_{10}\cdot\frac{\partial \overset{0\,\alpha}{V_b}}{\partial\nu}\Big|_S, \quad a_{10}(\tau) \neq 0. \tag{40}$$

Analogously, from the system (34) for $\kappa \geq 3$ we can determine recursion relations connecting the functions $\overset{\kappa\,i}{V_b}$ $(i = \tau, \alpha, \nu)$.

§ 3. Inclusion of the Boundary Conditions

Our problem is to choose vectors $\vec{u}_a$ and $\vec{u}_b$ satisfying the recursion relations (7)-(9), (17)-(19), (24)-(27), (29), (32), (33), (36), (37), and (40) such that the zero-stress boundary conditions (6) are satisfied on the surface S of the elastic body Ω in question.

In order for the boundary conditions to be tenable it is required that on S the vectors $\vec{u}_a$ and $\vec{u}_b$ have the same "phase," i.e., $\tau_a|_S = \tau_b|_S$ [see Eq. (12)]; otherwise $\vec{u}_a$ and $\vec{u}_b$ would have to satisfy the boundary conditions separately, and this would mean that $\vec{u}_a \equiv 0$ and $\vec{u}_b \equiv 0$.

The boundary conditions (6) ($i = \tau, \alpha, \nu$; order ω^1) yield the following, with regard for Eqs. (17), (19), (32), and (33):

$$\left.\begin{aligned} &2\tau_\nu^{\alpha} c^2 \overset{0}{V}{}_\alpha^{\tau} - c^2 l_1 \overset{0}{V}{}_b^{\nu} = 0, \\ &b^2 l_1 c^2 \overset{0}{V}{}_\alpha^{\tau} + 2\tau_\nu^{b} b^2 \overset{0}{V}{}_b^{\nu} = 0, \end{aligned}\right\} \tag{41}$$

$$\overset{0}{V}{}_b^{\alpha}\Big|_{\nu=0} = 0, \tag{42}$$

where

$$l_1 = \left(\frac{1}{b^2} - \frac{2}{c^2}\right).$$

By virtue of the Rayleigh equation

$$-4\tau_\nu^{q}\tau_\nu^{b} + l_1^2 c^2 = 0,$$

which holds on the entire surface S and, in particular, on the ray $\mathcal{L}$; the determinant of the system (41) is equal to zero.

The general solution of the system (41) has the form

$$\overset{0}{V}{}_\alpha^{\tau} = l_1 \chi_0(\tau,\eta), \qquad \overset{0}{V}{}_b^{\nu} = 2\tau_\nu^{\alpha} \chi_0(\tau,\eta), \tag{43}$$

where $\chi_0(\tau,\eta)$ is a new unknown function.

The boundary conditions in the higher approximations $\kappa = 2, 3, \ldots$ ($i = \tau, \alpha, \eta$; order $\omega^{1-\kappa/2}$) assume the form

$$\left.\begin{aligned} &2\tau_\nu^{\alpha} c^2 \overset{\kappa}{V}{}_\alpha^{\tau} - c^2 l_1 \overset{\kappa}{V}{}_b^{\nu} = \Lambda_1^{(\kappa)}(\overset{m}{V}{}_\alpha^{\tau}, \overset{m}{V}{}_b^{\nu}, \overset{m}{V}{}_b^{\alpha}), \\ &b^2 l_1 c^2 \overset{\kappa}{V}{}_\alpha^{\tau} + 2\tau_\nu^{b} b^2 \overset{\kappa}{V}{}_b^{\nu} = \Lambda_2^{(\kappa)}(\overset{m}{V}{}_\alpha^{\tau}, \overset{m}{V}{}_b^{\nu}, \overset{m}{V}{}_b^{\alpha}), \end{aligned}\right\} \tag{44}$$

$$\overset{\kappa}{V}{}_b^{\alpha} = \Lambda_3(\overset{m}{V}{}_\alpha^{\tau}, \overset{m}{V}{}_b^{\nu}, \overset{m}{V}{}_b^{\alpha}), \tag{45}$$

where $\Lambda_1^{(\kappa)}$, $\Lambda_2^{(\kappa)}$, and $\Lambda_3^{(\kappa)}$ are linear differential operators of at most second order relative to the functions $\overset{m}{V}{}_\alpha^{\tau}$, $\overset{m}{V}{}_b^{\nu}$, and $\overset{m}{V}{}_b^{\alpha}$, $m < \kappa$.

The necessary and sufficient condition for the existence of the functions $\overset{\kappa}{V}{}_\alpha^{\tau}$, $\overset{\kappa}{V}{}_b^{\nu}$ satisfying Eqs. (44) is written

$$\frac{2\tau_\nu^{b}}{c^2}\Lambda_1^{(\kappa)} + \frac{l_1}{b^2}\Lambda_2^{(\kappa)} = 0, \quad \kappa = 1, 2, \ldots. \tag{46}$$

For $\kappa = +1$ Eq. (46) holds, and the general solution of the system (44) has the form

$$\left.\begin{aligned} &\overset{1}{V}{}_\alpha^{\tau} = l_1 \chi_1(\tau,\eta) - \eta\left[\frac{\partial}{\partial\alpha}\ln(c^2\tau_\nu^{\alpha})\right] l_1 \chi_0(\tau,\eta), \\ &\overset{1}{V}{}_b^{\nu} = 2\tau_\nu^{\alpha}\chi_1(\tau,\eta) - \eta\left[\frac{\partial}{\partial\alpha}\ln(l_1 c^2)\right] 2\tau_\nu^{\alpha}\chi_0(\tau,\eta), \end{aligned}\right\} \tag{47}$$

where the second components of the functions $\overset{1}{V}{}_a^{\tau}$, and $\overset{1}{V}{}_b^{\nu}$ are particular solutions of (44) and $\chi_1(\tau, \eta)$ is a new unknown function.

For $\kappa = 2$ relation (46) may be written in the form

$$b_1 \frac{\partial \overset{0}{V}{}_a^{\tau}}{\partial \nu} + b_2 \frac{\partial \overset{0}{V}{}_b^{\nu}}{\partial \nu} + b_3 \frac{\partial \overset{0}{V}{}_a^{\tau}}{\partial \tau} + b_4 \frac{\partial \overset{0}{V}{}_b^{\nu}}{\partial \tau} + b_5 \frac{\partial^2 \overset{0}{V}{}_a^{\tau}}{\partial \eta^2} + b_6 \frac{\partial^2 \overset{0}{V}{}_b^{\nu}}{\partial \eta^2} + b_7 \overset{0}{V}{}_a^{\tau} + b_8 \overset{0}{V}{}_a^{\nu} = 0, \tag{48}$$

where $b_1, b_2, \ldots, b_8$ are functions of the variable τ.

Calculations show that Eq. (48) does not involve $\overset{1}{V}{}_a^{\tau}$, $\overset{1}{V}{}_b^{\nu}$, or the derivatives $\frac{\partial \overset{0}{V}{}_a^{\tau}}{\partial \eta}$, $\frac{\partial \overset{0}{V}{}_b^{\nu}}{\partial \eta}$.

Differentiation with respect to τ and η is the same as differentiation along the tangent to the surface S. Recognizing further that the coëfficients of Eqs. (43) are calculated at the point $(\tau, 0, 0)$, we rewrite Eq. (48) as follows:

$$b_1 \frac{\partial \overset{0}{V}{}_a^{\tau}}{\partial \nu} + b_2 \frac{\partial \overset{0}{V}{}_b^{\nu}}{\partial \nu} + (b_3 l_1 + 2 b_4 \tau_\nu^a) \frac{\partial \chi_0}{\partial \tau} +$$

$$+ (b_5 l_1 + 2 b_6 \tau_\nu^a) \frac{\partial^2 \chi_0}{\partial \eta^2} + \left(b_7 l_1 + 2\tau_\nu^a b_8 + b_3 \frac{\partial l_1}{\partial \tau} + 2 b_4 \frac{\partial \tau_\nu^a}{\partial \tau}\right) \chi_0 = 0. \tag{49}$$

Next we set $\nu = 0$ in Eqs. (26) and (38), whereupon, taking account of relations (43) and (49), we arrive at the following equation for the function χ after some rather lengthy intermediate calculations:

$$L \chi_0 + M \chi_0 = 0. \tag{50}$$

In Eq. (50) L is the linear differential operator

$$L = \frac{\partial^2}{\partial \eta^2} - K(\tau) \eta^2 + 2i \frac{\partial}{\partial \tau}, \tag{51}$$

$$K(\tau) = \frac{1}{2} \frac{\partial^2}{\partial \alpha^2} \left(\frac{c^2}{g_{\tau\tau}} \right) \bigg|_{\mathcal{L}}, \tag{52}$$

and $M = M(\tau)$ is a function representing a linear combination of the functions $\frac{\partial \lambda}{\partial \tau}$, $\frac{\partial \lambda}{\partial \alpha}$, $\frac{\partial \lambda}{\partial \nu}$, $\frac{\partial M}{\partial \tau}$, $\ldots$, $\frac{\partial \rho}{\partial \nu}$ calculated on $\mathcal{L}$.

We seek the solution of the parabolic equation (50) in the form (see [3])

$$\chi_0(\tau, \eta) = \chi_0(\tau) \exp\left(\frac{i}{2} \frac{\gamma'}{\gamma} \eta^2\right) \mathfrak{D}_q \left(\frac{\eta \sqrt{2}}{\gamma}\right), \tag{53}$$

where $\gamma = \gamma(\tau)$ and $\chi_0 = \chi_0(\tau)$ are unknown functions, γ has a real nonzero value, and $\mathfrak{D}_q$ are parabolic cylindrical functions [4] with integral index $q = 0, 1, 2, \ldots$, that satisfy the equation

$$\mathfrak{D}_q'' + \left[\left(q + \frac{1}{2}\right) - \frac{\eta^2}{2\gamma^2}\right] \mathfrak{D}_q = 0. \tag{54}$$

We now substitute a function $\chi_0(\tau, \eta)$ of the form (53) into Eq. (50). In order for the variables to be separable in the latter equation it suffices to have the functions $\gamma(\tau)$ and $\chi_0(\tau)$ obey the equations

$$\gamma'' + K(\tau)\gamma = \frac{1}{\gamma^3}, \tag{55}$$

$$\frac{d\chi_0}{d\tau} + \left(\frac{\gamma'}{2\gamma} + i\,\frac{q + 1/2}{\gamma^2}\right)\chi_0 + H\chi_0 = 0, \tag{56}$$

in which $K(\tau)$ is the quantity defined by Eq. (52) and the coefficient H coincides up to a constant multiplier with the coefficient K of Eq. (5.12) in [2], except only that H is calculated in the given instances on $\mathcal{L}$, i.e., with $\alpha = \nu = 0$.

Next we consider Eq. (55) (see the similar discussion in [3]).

Let $y_1(\tau)$ and $y_2(\tau)$ be two linearly-independent solutions of the equation

$$y'' + K(\tau)y = 0,$$

so that for the function $y(\tau)$ in the general case we have

$$y(\tau) = C_1 y_1(\tau) + C_2 y_2(\tau),$$

where C_1 and C_2 are arbitrary constants. We denote by $W[y_1, y_2]$ the Wronskian determinant of the linearly-independent solutions y_1 and y_2. Let us pick a second order symmetric matrix $\|d_{ij}\|$ such that

$$\det \|d_{ij}\| W^2[y_1, y_2] = 1.$$

In this case the function $\gamma(\tau)$ satisfying Eq. (55) corresponds to the formula

$$\gamma(\tau) = \left[\sum_{i,j=1}^{2} d_{ij}\, y_i(\tau)\, y_j(\tau)\right]^{1/2}. \tag{57}$$

We shall assume below that the functions $\gamma(\tau)$ comprising the solutions of Eq. (55) are known.

Let $\chi_0(\tau)$ be determined at some fixed point M_0 on $\mathcal{L}$, $(\chi_0(0))$. Equation (56) enables us to find $\chi_0(\tau)$ on $\mathcal{L}$ and Eq. (53) determines $\chi_0(\tau, \eta)$.

We call $\chi_0(\tau, \eta)$ the complex intensity of a Rayleigh wave concentrated near $\mathcal{L}$.

We now write the final form of the functions $\chi_0^{(q)}(\tau, \eta)$, corresponding to different values of the parameter q:

$$\chi_0^{(q)}(\tau, \eta) = \chi_0(0)\frac{1}{\sqrt[4]{g_{\alpha\alpha}}} \exp\int_0^\tau \frac{C}{A}\frac{b_{\tau\tau}}{c^2}\,d\tau \exp\int_0^\tau \frac{\mathfrak{D}}{A}\frac{b_{\alpha\alpha}}{g_{\alpha\alpha}}\,d\tau \exp\int_0^\tau \frac{E}{A}\,d\tau \exp\int_0^\tau \frac{F}{A}\,d\tau \times$$

$$\times \frac{1}{\sqrt{\gamma}}\,\frac{1}{g_{\tau\tau}} \exp\left[\frac{i}{2}\frac{\gamma'}{\gamma}\eta^2 - \frac{i}{2}\int_0^\tau (2q+1)\frac{d\tau}{\gamma^2}\right] \mathfrak{D}_q\left(\frac{\sqrt{2}\,\eta}{\gamma}\right), \tag{58}$$

$$q = 0, 1, 2, \ldots.$$

The coefficients A, C, $\mathfrak{D}$, E, and F in expression (58) coincide with the corresponding coefficients of (6.2) in [2], but here they are calculated on the ray $\mathcal{L}$, i.e., at the point $(\tau, (0, 0))$.

From relation (46) for $\kappa = 3$, we can find an equation of the form (50) for the function $\chi_1(\tau, \eta)$. Knowing the initial data of $\chi_1(\tau,\eta)$ at the point M_0 and taking account of the character of the sought-after solution (concentration near $\mathcal{L}$), we determine $\chi_1(\tau,\eta)$. Equations (47) make it possible from the known functions $\chi_0(\tau,\eta)$ and $\chi_1(\tau,\eta)$ to calculate $\overset{1}{V}{}_a^{\tau}$, $\overset{1}{V}{}_b^{\nu}$, and $\overset{1}{V}{}_b^{\alpha}$. Then, using the recursion relations of § 2, we proceed analogously to find the next higher coefficients $\overset{\kappa}{V}{}_j^{i}$ $(i = \tau, \alpha, \nu;\ j = a, b)$ for $\kappa \geq 2$ in the decompositions (14).

Calculating the components of the series (8) and (9) for the desired vector $\vec{U} = \vec{U}_a + \vec{U}_b$, we can formulate a solution satisfying the elastic-theory equations (2) and the boundary conditions (6) up to quantities of the order $\omega^{-N/2}$, where N is a given natural number.

§ 4. Conclusion

We are now ready to examine the asymptotic representation of the investigated solution.

The principal asymptotic term of the displacement vector $\vec{U}$ has the form

$$\vec{u} \sim \exp[-i\omega(t-\tau_a)]\,(\overset{0}{V}{}_a^{\tau}, \overset{0}{V}{}_a^{\alpha}, \overset{0}{V}{}_a^{\nu}) + \exp[-i\omega(t-\tau_b)]\,(\overset{0}{V}{}_b^{\tau}, \overset{0}{V}{}_b^{\alpha}, \overset{0}{V}{}_b^{\nu}), \tag{59}$$

where

$$\begin{gathered}
\overset{0}{V}{}_a^{\tau} = \ell_1 \chi_0, \quad \overset{0}{V}{}_a^{\alpha} = 0, \quad \overset{0}{V}{}_a^{\nu} = c^2 \tau_\nu^a \ell_1 \chi_0, \\
\overset{0}{V}{}_b^{\tau} = -2\tau_\nu^a \tau_\nu^b \chi_0, \quad \overset{0}{V}{}_b^{\alpha}\Big|_S = 0, \quad \overset{0}{V}{}_b^{\nu} = 2\tau_\nu^a \chi_0, \\
\frac{\partial \overset{0}{V}{}_b^{\alpha}}{\partial \nu}\Bigg|_{\nu=0} \neq 0,
\end{gathered} \tag{60}$$

and the function $\chi_0 = \chi_0(\tau, \eta)$ is calculated from Eq. (58).

I°. Inasmuch as the limit of the parabolic cylindrical functions $\mathcal{D}_q(z)$ as $z \to \pm\infty$ is equal to zero only for nonnegative integral values of q, we can assume that $q = 0, 1, 2, \ldots$. The functions $\mathcal{D}_q(z)$ oscillate if

$$\sqrt{2q+1} > \left| \frac{\sqrt{\omega}\,\alpha}{\gamma} \right|, \qquad z = \frac{\sqrt{2}}{\gamma}\eta,$$

and decay exponentially if

$$\sqrt{2q+1} < \left| \frac{\sqrt{\omega}\,\alpha}{\gamma} \right|.$$

Thus, for a sufficiently high frequency ω the solution (stopping with the zeroth approximation of the displacement vector $\vec{u}$)

$$\begin{aligned}
\vec{u}^{(q)} &\sim \exp(-i\omega t)\Big\{[\exp(i\omega\tau_a)]\,\ell_1\,(1, 0, c^2\tau_\nu^a) + \\
&+ [\exp(i\omega\tau_b)]\, 2\tau_\nu^a\,(-\tau_\nu^b, 0, 1)\Big\}\,\hat{\chi}_0 \frac{1}{\sqrt{\gamma}} \times \\
&\times \exp\left[\frac{i}{2}\frac{\gamma'}{\gamma}\eta^2 - \frac{i}{2}\int_0^{\tau}(2q+1)\frac{d\tau}{\gamma^2}\right]\frac{1}{c^2}\,\mathcal{D}_q\!\left(\frac{\eta\sqrt{2}}{\gamma}\right),
\end{aligned} \tag{61}$$

in which

$$\tau_\nu^a = \frac{\partial \tau_a}{\partial \nu}\bigg|_s, \quad \tau_\nu^b = \frac{\partial \tau_b}{\partial \nu}\bigg|_s,$$

$$\tau_a = \tau + i\nu \sqrt{\frac{1}{c^2} - \frac{1}{a^2}}\bigg|_{\mathcal{L}} + i\nu\alpha \frac{\partial}{\partial \alpha}\sqrt{\frac{1}{c^2} - \frac{1}{a^2}}\bigg|_{\mathcal{L}} + \ldots,$$

$$\tau_b = \tau + i\nu \sqrt{\frac{1}{c^2} - \frac{1}{b^2}}\bigg|_{\mathcal{L}} + i\nu\alpha \frac{\partial}{\partial \alpha}\sqrt{\frac{1}{c^2} - \frac{1}{b^2}}\bigg|_{\mathcal{L}} + \ldots,$$

$$\hat{\chi}_0 = \chi_0(0)\frac{1}{\sqrt[4]{g_{\alpha\alpha}}} \exp \int_0^\tau \frac{C}{A}\frac{b_{\tau\tau}}{c^2}\,d\tau \exp\int_0^\tau \frac{D}{A}\frac{b_{\alpha\alpha}}{g_{\alpha\alpha}}\,d\tau \exp\int_0^\tau \frac{E}{A}\,d\tau \exp\int_0^\tau \frac{F}{A}\,d\tau,$$

differs significantly from zero only in the narrow strip

$$|\alpha| \leqslant \frac{\sqrt{2q+1}\,\gamma}{\sqrt{\omega}}, \quad (\omega \to \infty), \tag{62}$$

about the ray $\mathcal{L}$. Outside the limits of this strip the solution decays exponentially. The thickness of the strip (62) of concentration of the solution is quantized ($q = 0, 1, 2, \ldots$).

2°. The Rayleigh waves thus constructed are essentially surface modes. The moduli of both components in expression (61) may be written in the form

$$|\vec{u}_a| = O(\omega^0)\exp\left(-\nu\omega\sqrt{\frac{1}{c^2} - \frac{1}{a^2}}\right),$$

$$|\vec{u}_b| = O(\omega^0)\exp\left(-\nu\omega\sqrt{\frac{1}{c^2} - \frac{1}{b^2}}\right).$$

Consequently, the displacement vector $\vec{u}$, together with the complex longitudinal wave $\vec{u}_a$ and the complex transverse wave $\vec{u}_b$ decay exponentially with depth, tending to zero as $\nu \to +\infty$. The higher the frequency ω, the faster will be the decay. As $\omega \to \infty$ the moduli of $\vec{u}_a$ and $\vec{u}_b$ will differ significantly from zero in the boundary strip $\nu = O\left(\frac{1}{\omega}\right)$, and for $\nu = o\left(\frac{1}{\omega^\alpha}\right), \alpha < 1$, it will tend to zero exponentially.

3°. Let us compare the asymptotic representations of the generalized steady-state Rayleigh waves found in [2] by the ray method and in the present article by the parabolic equation method. We recall that the steady-state harmonic Rayleigh waves of [2] were obtained from nonsteady waves [see Eqs. (2.1) and (6.1) in that article] for

$$\vec{u} = \sum_{k=0}^{\infty} \vec{u}_k(\tau, \alpha, \nu) f_k(t - \tau), \quad f_k = \exp[-i\omega(t-\tau)](i\omega)^{-k}.$$

The asymptotic representation of the displacement vector $\vec{u}$ for the Rayleigh surface waves investigated in [2] with regard for the harmonic character of these waves is given by Eqs. (6.1) and (6.4), which are presented herewith for convenience:

$$\vec{u} \sim l_1 \chi_0(\tau, \alpha)\, \nabla\tau_a \exp[-i\omega(t - \tau_a)] + l_2 \chi_0(\tau, \alpha)\,\vec{\xi} \exp[-i\omega(t - \tau_b)], \tag{63}$$

$$\chi_0(\tau,\alpha) = \chi_0(0,\alpha)\frac{1}{\sqrt[4]{g_{\alpha\alpha}}}\int_0^\tau \frac{C}{A}\frac{b_{\tau\tau}}{c^2}\,d\tau\,\exp\int_0^\tau \frac{D}{A}\frac{b_{\alpha\alpha}}{g_{\alpha\alpha}}\,d\tau\,\exp\int_0^\tau \frac{E}{A}\,d\tau\int_0^\tau \frac{F}{A}\,d\tau, \tag{64}$$

where

$$\left.\begin{aligned} & l_1 = \left(\frac{1}{b^2}-\frac{2}{c^2}\right), \quad l_2 = \frac{2i}{c}\sqrt{\frac{1}{c^2}-\frac{1}{a^2}}, \\ & \vec{\xi} = -\frac{1}{b}\frac{1}{\sqrt{1-b^2(\tau_\nu^b)^2}}\left(\vec{n}-b^2\tau_\nu^b\nabla\tau_b\right). \end{aligned}\right\} \tag{65}$$

For Rayleigh waves concentrated near $\mathcal{L}$ we have from Eqs. (58) and (61)

$$\vec{u} \sim l_1\chi_0(\tau,\eta)c^2\nabla\tau_a\exp\left[-i\omega(t-\tau_a)\right] + l_2\chi_0(\tau,\eta)c^2\vec{\xi}\exp\left[-i\omega(t-\tau_b)\right], \tag{66}$$

$$\begin{aligned} \chi_0(\tau,\eta) = {} & \chi_0(0)\frac{1}{\sqrt[4]{g_{\alpha\alpha}}}\exp\int_0^\tau \frac{C}{A}\frac{b_{\tau\tau}}{c^2}\,d\tau \times \\ & \times\exp\int_0^\tau \frac{D}{A}\frac{b_{\alpha\alpha}}{g_{\alpha\alpha}}\,d\tau\,\exp\int_0^\tau \frac{E}{A}\,d\tau\,\exp\int_0^\tau \frac{F}{A}\,d\tau \times \frac{1}{c^2}\times \\ & \times\frac{1}{\sqrt{\gamma}}\exp\left[\frac{i}{2}\frac{\gamma'}{\gamma}\eta^2 - \frac{i}{2}\int_0^\tau \frac{2q+1}{\gamma^2}\,d\tau\right]D_q\left(\frac{\sqrt{2}\,\eta}{\gamma}\right), \quad \alpha\omega^{1/2} = \eta. \end{aligned} \tag{67}$$

An analysis of Eqs. (63), (64) and (66), (67) shows that they are consistent. The displacement vectors as functions of τ and ν describe identical wave processes. With respect to the variable α their behavior differs, as in the first case the Rayleigh waves propagate over the entire surface, while in the second case they are concentrated near the ray $\mathcal{L}$ on the surface.

As to be expected, the concentrated waves turn out to be a special case of the surface waves of [2]. However, the ray method used in [2] does not allow the concentrated solutions to be formulated directly.

The author conveys his deepest appreciation to Prof. V. M. Babich for valuable consultation and continuing aid in the study.

LITERATURE CITED

1. Babich, V. M., Propagation of Rayleigh waves over the surface of a homogeneous elastic body of arbitrary shape, Dokl. Akad. Nauk SSSR, Vol. 137, No. 6 (1961).
2. Babich, V. M. and Rusakova (Kirpichnikova), N. Ya., Propagation of Rayleigh waves over the surface of an inhomogeneous elastic body of arbitrary shape, Zh. Vychislit. Mat. i Mat. Fiz., Vol. 2, No. 4 (1962).
3. Babich, V. M. and Lazutkin, V. F., Eigenfunctions concentrated near a closed geodesic, in: Topics in Mathematical Physics, Vol. 2, Consultants Bureau, New York (1968), pp. 9-18.
4. Whittaker, E. T. and Watson, G. N., Modern Analysis, Vol. 2, Cambridge Univ. Press (1927).

ESTIMATION OF THE INTENSITY OF RAYLEIGH AND STONELY SURFACE WAVES ON AN INHOMOGENEOUS PATH

P. V. Krauklis

At even relatively short distances from the source of ground and underwater explosions the greater part of the elastic energy is given over to surface waves, a fact that stems primarily from the weaker damping of surface waves relative to volume waves. Often for the assessement of the intensity of surface waves the linear theory of elasticity is used, within the framework of which it is possible to describe the fundamental qualitative characteristics of the wave pattern. The most widely used model for theoretical investigations is a homogeneous half-space or a layer of water on a homogeneous half-space. However, the problem of the influence of ground inhomogeneity on such characteristics of surface waves as the intensity and wave form has not yet been solved. In connection with the publication of [1], in which a method is proposed for determining the energy conservation laws for Rayleigh surface waves, it has proved possible to solve the problem on the basis of the ray method. In [2] the results of [1] were extended to the case of Stonely waves generated at the interface between a liquid and solid or between two solids. The resulting formulas enable one to account for the geometry of the surface over which the wave is traveling as well as the elastic parameters of the ground along the path.

In the present article we apply the results of [1, 2] to the problem of a finite source of the ground or underwater explosion type. We first examine in detail Rayleigh waves on the free boundary of an inhomogeneous half-space, then concisely formulate the results for Stonely waves at the interface between a liquid and a solid half-space.

Consider an inhomogeneous body bounded by an arbitrary smooth surface S, to which is applied at a point M_0 a point source of the normal force type:

$$\vec{t} = \frac{\delta(\tau)}{\tau}\,\delta(t)\,\vec{n}. \tag{1}$$

The vector $\vec{n}$ is a unit vector directed along the normal to S and inward toward the body. We introduce a curvilinear coordinate system α, τ, ν, where α and τ are the ray coordinates on the surface and ν is the distance along the normal. From the point M_0 along S we draw rays extremizing the interval $\int_{M_0}^{M} \frac{ds}{c}$ (where ds is an arc element, c is the local Rayleigh velocity, and M is any point on the surface). Every ray is characterized by the parameter α, and every point on a ray is characterized by the parameter τ, which is equal to the traversal time of the wave from M_0 to M. As shown in [3], the Rayleigh wave displacement vector can be represented in the form

$$\vec{u} = e_1 \chi_0 f(t-\tau_a)\nabla\tau_a + e_2 \chi_0 f(t-\tau_b)\vec{\xi}, \tag{2}$$

where

$$e_1=\frac{1}{b^2}-\frac{2}{c^2},\quad e_2=\frac{2}{c}\,i\sqrt{\frac{1}{c^2}-\frac{1}{a^2}},\quad \tau_a=\tau-i\nu\sqrt{\frac{1}{c^2}-\frac{1}{a^2}},$$

$$\tau_b=\tau-i\nu\sqrt{\frac{1}{c^2}-\frac{1}{b^2}},\quad \vec{\xi}=\frac{-1}{b\sqrt{1-b^2\left(\frac{\partial\tau_b}{\partial\nu}\right)^2}}\cdot\left(\vec{n}-b^2\frac{\partial\tau_b}{\partial\nu}\nabla\tau_b\right). \tag{3}$$

In Eqs. (2)-(3) a and b denote the local longitudinal and transverse wave velocities, respectively, and the function χ_0 characterizes the Rayleigh wave intensity. It follows from the integral law of energy conservation that the following expression must be left invariant along a ray:

$$|\chi_0|^2|X_\alpha|\frac{\rho}{c\sqrt{\frac{1}{c^2}-\frac{1}{b^2}}}\left[\frac{6}{c^2}\left(\frac{1}{b^2}-\frac{1}{a^2}\right)+\frac{4}{a^2b^2}-\frac{6}{b^4}+\frac{c^2}{b^6}\right]=\text{const.} \tag{4}$$

In Eq. (4) the function $\sqrt{|X_\alpha|}$ is the geometric divergence of the rays and ρ is the local density of the ground. The description of the source and Eqs. (3) and (4) make it possible to determine the Rayleigh wave intensity on an inhomogeneous path. The form of the function χ_0 near the source is found from the solution of the corresponding homogeneous problem as follows:

We consider a homogeneous half-space $z>0$, to the boundary of which at the point $r=0$, $z=0$ is applied the normal force given by Eq. (1). Representing the potential in the form

$$\varphi=\frac{1}{2\pi i}\int_0^\infty J_0(\kappa r)\,d\kappa\int_{\sigma-i\infty}^{\sigma+i\infty}A\exp\kappa[t\eta b-z\alpha]\,d\eta,$$

$$\psi=\frac{1}{2\pi i}\int_0^\infty J_1(\kappa r)\,d\kappa\int_{\sigma-i\infty}^{\sigma+i\infty}B\exp\kappa[t\eta b-z\beta]\,d\eta, \tag{5}$$

and invoking the standard formulas relating the displacements and stresses to the potentials, as well as the boundary conditions for the determination of A and B, we obtain the following expressions for the horizontal and vertical components of the Rayleigh wave displacements:

$$q=-\frac{b}{2\tau\mu D}\,\mathrm{Re}\int_0^\infty i\kappa J_1(\kappa r)\left(-ge^{-\kappa\alpha z}+2\alpha\beta e^{-\kappa\beta z}\right)e^{i\kappa ct}\,d\kappa,$$

$$w=-\frac{b\alpha}{2\tau\mu D}\,\mathrm{Re}\int_0^\infty i\kappa J_0(\kappa r)\left(-ge^{-\kappa\alpha z}+2e^{-\kappa\beta z}\right)e^{i\kappa c t}\,d\kappa, \tag{6}$$

$$\tau=\frac{c}{b},\quad g=2-\tau^2,\quad \alpha=\sqrt{1-\gamma^2\tau^2},\quad \beta=\sqrt{1-\tau^2},$$

$$\gamma=\frac{b}{a},\quad \mu=b^2\rho,\quad D=g-\frac{\alpha}{\beta}-\frac{\gamma^2\beta}{\alpha}.$$

If we use the asymptotic representation of the Bessel functions and change to the new variable $\omega=\kappa c$ in the integral of (6), for q and w we obtain the final expressions

$$q = \frac{\sqrt{2}}{4} \frac{b^2}{\mu \sqrt{\pi r}\, D c^2 \sqrt{c}} \operatorname{Re} \int_0^\infty \sqrt{\omega} \left(-g e^{-\frac{\omega}{c}\alpha z} + 2\alpha\beta e^{-\frac{\omega}{c}\beta z} \right) e^{i\omega\left(t - \frac{r}{c}\right) + i\frac{\pi}{4}} \, d\omega,$$

$$w = \frac{\sqrt{2}}{4} \frac{b}{\mu \sqrt{\pi r}\, D c^2 \sqrt{c}} \operatorname{Re} \int_0^\infty i\sqrt{\omega}\, \alpha \left(-g e^{-\frac{\omega}{c}\alpha z} + 2 e^{-\frac{\omega}{c}\beta z} \right) e^{i\omega\left(t - \frac{r}{c}\right) + i\frac{\pi}{4}} \, d\omega. \tag{7}$$

Comparing (7) and (2), we obtain an explicit expression for χ_0:

$$\chi_0 = \frac{\sqrt{2}}{4} \frac{b^2 \sqrt{c}}{\mu \sqrt{\pi r}\, D} = \frac{\sqrt{2}}{4} \frac{1}{D\rho} \sqrt{\frac{c}{\pi r}}. \tag{8}$$

Consequently, χ_0 in (8) characterizes the variation of the intensity for Rayleigh waves in the case of an inhomogeneous medium. Now, beginning with $r = r_1$, let the parameters of the medium begin to vary both in depth and along the path. In this case, as indicated in [3], the variation of the elastic properties with depth only affects the waveform [i.e., $f(t - \tau_a)$ and $f(t - \tau_b)$], while the variation along the path changes the wave amplitude. Equation (4) permits this variation to be accounted for. The following relation applies to the ratio of the amplitudes at points (1) and (2):

$$\frac{J_{2w}}{J_{1w}} = \sqrt{\frac{|X\alpha_1|}{|X\alpha_2|}} \frac{\tau_2 \sqrt{1-\gamma_2^2\tau_2^2}\, \sqrt[4]{1-\tau_2^2}\, \sqrt{\rho_1}\, c_1 \left[6(1-\gamma_1^2) + 4\gamma_1^2\tau_1^2 - 6\tau_1^2 + \tau_1^4\right]^{1/2}}{\tau_1 \sqrt{1-\gamma_1^2\tau_1^2}\, \sqrt[4]{1-\tau_1^2}\, \sqrt{\rho_2}\, c_2 \left[6(1-\gamma_2^2) + 4\gamma_2^2\tau_2^2 - 6\tau_2^2 + \tau_2^4\right]^{1/2}}, \tag{9}$$

$$\frac{J_{2q}}{J_{1q}} = \sqrt{\frac{|X\alpha_1|}{|X\alpha_2|}} \frac{\tau_1 \left(2 - \tau_2^2 - 2\sqrt{1-\gamma_2^2\tau_2^2}\, \sqrt{1-\tau_2^2}\right) \sqrt{\rho_1}\, \sqrt[4]{1-\tau_2^2}\, c_1 \left[6(1-\gamma_1^2) + 4\gamma_1^2\tau_1^2 - 6\tau_1^2 + \tau_1^4\right]^{1/2}}{\tau_2 \left(2 - \tau_1^2 - 2\sqrt{1-\gamma_1^2\tau_1^2}\, \sqrt{1-\tau_1^2}\right) \sqrt{\rho_2}\, \sqrt[4]{1-\tau_1^2}\, c_2 \left[6(1-\gamma_2^2) + 4\gamma_2^2\tau_2^2 - 6\tau_2^2 + \tau_2^4\right]^{1/2}}. \tag{10}$$

For example, let a Rayleigh wave travel from a ground with parameters $a_1 = 5000$ μ/sec, $b_1 = 2885$ μ/sec, and $\rho_1 = 2$ g/cm³ to a ground in which $a_2 = 3000$ μ/sec, $b_2 = 1200$ μ/sec, and $\rho_2 = 2$ g/cm³ so that, accordingly, $\gamma_1 = 0.577$, $\tau_1 = 0.919$, $\gamma_2 = 0.4$, and $\tau_2 = 0.943$. We neglect the amplitude variation due to geometric divergence. Then from Eqs. (9) and (10) we obtain

$$\frac{J_{2w}}{J_{1w}} = 2.06, \qquad \frac{J_{2q}}{J_{1q}} = 1.81. \tag{11}$$

It is clear that all of the foregoing arguments carry over to Rayleigh waves generated by a shock wave. The form of the function χ_0 is determined in this case as above for a stationary source.

Our next concern is the determination of the intensity of Stonely waves from an underwater explosion. The energy conservation law derived for these waves in [2] can be written in the form of the equation

$$|\chi_0|^2 |X_a| c \left\{ \rho_1 \left[\frac{g_1}{\alpha_1 c^2}\left(g_1 - 8\alpha_1^2\right) + \frac{4\alpha_1^2}{\beta_1 c^2}\left(\frac{3}{c^2} - \frac{2}{b_1^2}\right) + \frac{\rho_0 \alpha_1^2}{c^2 \alpha_0^3 b_1^2 a_0^2} \right\} = \text{const}. \tag{12}$$

In Eq. (12) the indices 1 and 0 refer to the ground and the water, respectively, where $g_1 = \frac{2}{c^2} - \frac{1}{b_1^2}$, $\alpha_0 = \sqrt{\frac{1}{c^2} - \frac{1}{a_0^2}}$, $\alpha_1 = \sqrt{\frac{1}{c^2} - \frac{1}{a_1^2}}$, $\beta_1 = \sqrt{\frac{1}{c^2} - \frac{1}{b_1^2}}$. The Stonely wave velocity c involved in (12) is determined from the equation

$$\alpha_0\left[\left(\frac{2}{c^2}-\frac{1}{b_1^2}\right)^2-\frac{4}{c^2}\alpha_1\beta_1\right]+\frac{\rho_0}{\rho_1}\frac{\alpha_1}{b_1^4}=0 \tag{13}$$

and is a function of $\gamma_1=\frac{b_1}{a_1}$, $\gamma_0=\frac{b_1}{a_0}$, and $\rho_{01}=\frac{\rho_0}{\rho_1}$. The Stonely wave displacement vectors in the liquid and solid are equal to

$$u_{a0}=\varphi_0 f(t-\tau_{a0})\nabla\tau_{a0}, \quad u_{a1}=\varphi_1 f(t-\tau_{a1})\nabla\tau_{a1},$$

$$u_{b1}=\psi_1 f(t-\tau_{b1})\vec{\xi},$$

where

$$\tau_{a1}=\tau+i\alpha_1\nu, \quad \tau_{b1}=\tau+i\beta_1\nu, \quad \tau_{a0}=\tau-i\nu\alpha_0,$$

$$\varphi_0=e_3\chi_0, \quad \varphi_1=e_1\chi_0, \quad \psi_1=e_2\chi_0, \tag{14}$$

$$e_1=\frac{1}{b_1^2}-\frac{2}{c^2}, \quad e_2=\frac{2}{c}i\sqrt{\frac{1}{c^2}-\frac{1}{a_1^2}}, \quad e_3=\frac{\alpha_1}{\alpha_0 b_1^2}.$$

Numerical calculations show that for $\gamma_1=0.4$, $\gamma_0=1.2$, and $\rho_{01}=0.5$ the velocity c is equal to $0.77b_1$; for $\gamma_1=0.5$, $\gamma_0=1.5$, and $\rho_{01}=0.5$ the velocity $c=0.65\,b_1$; and for $\gamma_1=0.577$, $\gamma_0=1.731$, and $\rho_{01}=0.5$ the velocity $c=0.570\,b_1$. The wave displacement amplitudes in water, calculated for the case of wave transmission from a ground with parameters $\gamma_1=0.577$, $\gamma_0=1.730$, $\rho_{01}=0.5$, $b_1=2600\,\mu$/sec, and $c=1480\,\mu$/sec to a ground with $\gamma_1=0.4$, $\gamma_0=1.2$, $\rho_{01}=0.5$, $b_1=1800\,\mu$/sec, $c=1380\,\mu$/sec, turn out to be equal to

$$\frac{u_{q_2}}{u_{q_1}}=1.86 \qquad \frac{u_{w_2}}{u_{w_1}}=4.13. \tag{15}$$

The calculations exhibit a considerable dependence of the displacement amplitudes in the water wave on the elastic characteristics of the bottom. It is important to note in conclusion that the proposed techniques for determining the displacement amplitudes of surface waves are justified for the case of sudden disturbances such as point detonations.

LITERATURE CITED

1. Babich, V. M., On energy conservation in the propagation of nonstationary waves, Vest. Leningrad. Univ., No. 7 (1967).
2. Krauklis, P. V., Nonstationary Stonely waves, in: Problems in the Dynamical Theory of Wave Propogation, No. 9 (1968).
3. Babich, V. M., and Rusakova, N. Ya., Propagation of Rayleigh waves over the surface of an inhomogeneous elastic body of arbitrary shape, Vychislit. Mat. i Mat. Fiz., Vol. 2, No. 4 (1962).

EIGENMODES IN AN ANNULAR RESONATOR: THE VECTORIAL PROBLEM

T. F. Pankratova

We propose to solve the system of Maxwell equations for a three-dimensional multimirror annular resonator filled with an inhomogeneous isotropic medium. We use the parabolic equation method to formulate the high-frequency asymptotic behavior of the eigenvalues and eigenfunctions concentrated in the vicinity of the resonator axis (see [1]). We find proper corrections to the eigenfrequencies and recursion formulas for any approximation. We show that the problem is not reducible to a scalar problem in the general case.

§ 1. Statement of the Problem

Consider an N-mirror resonator filled with an inhomogeneous medium characterized by a dielectric constant $\varepsilon(\vec{r})$ and permeability $\mu(\vec{r})$, where $\vec{r}$ is the radius vector of a point in three-dimensional space. Regarding all processes as linear and undamped ($\operatorname{Im}\varepsilon = \operatorname{Im}\mu = 0$), we write the field equations:

$$\begin{cases} \operatorname{rot} \vec{E} = i\omega\mu \vec{H}, \\ \operatorname{rot} \vec{H} = -i\omega\varepsilon \vec{E}. \end{cases} \tag{1.1}$$

Consistent with the adopted terminology, we define rays as the extremals of the functional $\mathcal{J} = \int_{M_0}^{M} \frac{1}{\vartheta(r)}\, d\sigma$, where $d\sigma$ is an element of length and $\vartheta = \frac{1}{\sqrt{\varepsilon\mu}}$ is the wave propagation velocity. Let the mirrors be arranged so that there is a closed ray (resonator axis) that goes into itself after reflection. We call the segment of the axis between the $(j-1)$st and j-th mirrors the j-th arm of the resonator.

We seek solutions to (1.1) concentrated in the vicinity of the resonator axis, i.e., we require

$$\max\{|E|, |H|\} \xrightarrow[\sqrt{\omega}R \to \infty]{} 0, \tag{1.2}$$

where R is the distance from the resonator axis.

Let us consider the case of ideal reflection, when the following boundary conditions are met on the surface of the mirrors:

$$E_{t_i}\Big|_{S_j} = 0, \qquad i = 1, 2, \qquad j = 1, 2, \ldots, N, \tag{1.3}$$

where E_{t_i} are the components of the vector $\vec{E}$, tangent to the reflecting surface.

We require that waves propagating along the resonator axis close:

$$\vec{E}(s+L) = \vec{E}(s),$$
$$\vec{H}(s+L) = \vec{H}(s), \tag{1.4}$$

where L is the length of the entire axis.

We approximately formulate solutions to the system (1.1) in each arm of the resonator, relating them to one another by conditions (1.3), and from the closure requirement (1.4) we obtain the eigen-frequencies of the resonator.

§ 2. Formulation of Solutions of the System (1.1) by the Parabolic Equation Method; Zeroth Approximation

Following [2], we introduce in the vicinity of each arm of the resonator a coordinate system (s_j, ξ_j, η_j) such that the radius vector of any point M has the form $\vec{r}(M) = \vec{r}(s) + \xi\vec{\ell}_\xi + \eta\vec{\ell}_\eta$, where $\vec{r}(s)$ is the radius vector of a point on the geodesic, the unit vectors $\vec{\ell}_\xi$ and $\vec{\ell}_\eta$ lie in the plane perpendicular to the ray and are rotated about the normal and binormal of the curve through an angle $\theta = \int_{s_0}^{s} \varkappa ds + \theta_0$, and $\varkappa(s)$ is the torsion of the curve. The coordinate system is orthogonal, and its Lamé coefficients are equal to

$$h_\xi = h_\eta = 1, \quad h_s = 1 - R(s)(\xi\cos\theta + \eta\sin\theta),$$

where $R(s)$ is the curvature of the curve.

We now write the Maxwell equations (1.1) in the coordinates (s, ξ, η). We seek the solution in the form

$$\vec{E} = \frac{1}{\sqrt{\varepsilon}}\vec{\mathcal{E}}(s,\xi,\eta)e^{i\omega\tau}, \qquad \vec{\mathcal{E}} = \sum_{k=0}^{\infty}\vec{\mathcal{E}}^{k}\omega^{-\frac{k}{2}},$$

$$\vec{H} = \frac{1}{\sqrt{\mu}}\vec{\mathcal{H}}(s,\xi,\eta)e^{i\omega\tau}, \qquad \vec{\mathcal{H}} = \sum_{k=0}^{\infty}\vec{\mathcal{H}}^{k}\omega^{-\frac{k}{2}},$$

where $\vec{\mathcal{E}}$ and $\vec{\mathcal{H}}$ are the Fok attenuation factors and τ is the eikonal: $\tau = \int_0^s \frac{1}{\vartheta(s,0,0)} ds$. In accordance with the parabolic equation method we assume

$$\left.\begin{aligned} &\omega >> 1, \\ &\xi, \eta \sim o(\omega^{-\frac{1}{2}}), \\ &A_i = o(1), \\ &\frac{\partial^{p+q}}{\partial\xi^p\partial\eta^q}A_i = O(\omega^{\frac{p+q}{2}}), \end{aligned}\right\} \tag{2.3}$$

where A_i is any component of the vectors $\vec{E}$ or $\vec{H}$.

All other variables are on the order of unity.

Let us decompose all the functions involved in Eq. (1.1) in a Taylor series in the vicinity of the geodesic $\xi = \eta = 0$. We introduce the new variables $x = \sqrt{\omega}\,\xi$ and $y = \sqrt{\omega}\,\eta$, where $x, y \sim o(1)$.

Setting the coefficients of different powers of ω equal to zero and recognizing that the axis $(s, 0, 0)$ is an extremal of the functional $\mathfrak{J}$, we obtain the following relations for the components of the vectors $\vec{\mathcal{E}}^{\kappa}$ and $\vec{\mathcal{H}}^{\kappa}$:

For the longitudinal components

$$
\begin{aligned}
&\overset{0}{\mathcal{E}}_s = \overset{0}{\mathcal{H}}_s = 0,\\
&\overset{1}{\mathcal{E}}_s = i\vartheta(s,0,0)\left(\frac{\partial}{\partial x}\overset{0}{\mathcal{H}}_{\eta} - \frac{\partial}{\partial y}\overset{0}{\mathcal{H}}_{\xi}\right),\\
&\overset{1}{\mathcal{H}}_s = -i\vartheta(s,0,0)\left(\frac{\partial}{\partial x}\overset{0}{\mathcal{E}}_{\eta} - \frac{\partial}{\partial y}\overset{0}{\mathcal{E}}_{\xi}\right),\\
&\overset{2}{\mathcal{E}}_s = i\vartheta(s,0,0)\left(\frac{\partial}{\partial x}\overset{1}{\mathcal{H}}_{\eta} - \frac{\partial}{\partial y}\overset{1}{\mathcal{H}}_{\xi}\right) - \frac{i\vartheta(s,0,0)}{2}\left(\left(\frac{1}{\mu}\frac{\partial \mu}{\partial \xi}\right)_s \overset{0}{\mathcal{H}}_{\xi} - \frac{1}{\mu}\frac{\partial \mu}{\partial \eta}\Big|_s \overset{0}{\mathcal{H}}_{\eta}\right),\\
&\overset{2}{\mathcal{H}}_s = -\vartheta(s,0,0)\left(\frac{\partial}{\partial x}\overset{1}{\mathcal{E}}_{\eta} - \frac{\partial}{\partial y}\overset{1}{\mathcal{E}}_{\xi}\right) + \frac{i\vartheta(s,0,0)}{2}\left(\frac{1}{\varepsilon}\frac{\partial \varepsilon}{\partial \xi}\Big|_s \overset{0}{\mathcal{E}}_{\xi} - \frac{1}{\varepsilon}\frac{d\varepsilon}{\partial \eta}\Big|_s \overset{0}{\mathcal{E}}_{\eta}\right),
\end{aligned}
\tag{2.4}
$$

etc.

Equations of the type (2.4) determine the longitudinal components $\overset{\kappa}{\mathcal{E}}_s$ and $\overset{\kappa}{\mathcal{H}}_s$ for any κ in terms of the functions $\overset{m}{\mathcal{E}}_{\xi}$, $\overset{m}{\mathcal{E}}_{\eta}$, $\overset{m}{\mathcal{H}}_{\xi}$, and $\overset{m}{\mathcal{H}}_{\eta}$ and their derivatives with respect to x and y for $m \leq \kappa - 1$. Thus we have reduced the problem to that of merely finding the transverse components, which form a one-column matrix:

$$
V^{\kappa} = \begin{pmatrix} \overset{\kappa}{\mathcal{E}}_{\xi} \\ \overset{\kappa}{\mathcal{E}}_{\eta} \\ \overset{\kappa}{\mathcal{H}}_{\xi} \\ \overset{\kappa}{\mathcal{H}}_{\eta} \end{pmatrix}
\tag{2.5}
$$

For the transverse components

$$
AV^{0} = 0, \quad AV^{1} = 0, \quad AV^{2} = BV^{0}, \ldots, \tag{2.6}
$$

$$
A = \begin{pmatrix} 0 & 1 & 1 & 0 \\ 1 & 0 & 0 & -1 \\ -1 & 0 & 0 & 1 \\ 0 & -1 & -1 & 0 \end{pmatrix}, \quad \det A = 0,
$$

where B is a 4×4 matrix, the elements of which involve the differentiation operators with respect to s, x, and y.

For each of the functions

$$
\overset{0}{\varphi}_i = \frac{1}{\sqrt{\vartheta(s)}}\overset{0}{\mathcal{E}}_i, \quad i = \xi, \eta, \tag{2.7}
$$

the stipulation of inhomogeneous solvability of the system (2.6) yields the customary parabolic equation

$$
L\overset{0}{\varphi}_i = 0, \tag{2.8}
$$

where

$$L \equiv \frac{\partial}{\partial x^2} + \frac{\partial}{\partial y^2} + 2\frac{i}{\vartheta(s)}\frac{\partial}{\partial s} - \frac{1}{\vartheta^2(s)}(\mathcal{K}(s)\vec{\varkappa}, \vec{\varkappa}),$$

$$\mathcal{K}(s) = \frac{1}{\vartheta(s)} \left\| \frac{\partial \vartheta}{\partial q_i \partial q_k}\Big|_s \right\|, \qquad q_i = \xi, \eta, \quad q_k = \xi, \eta, \qquad \vec{\varkappa} = (x, y).$$

Equation (2.8) is satisfied by the function (see [1])

$$\overset{\circ}{\varphi}_i = \frac{c_i}{\sqrt{\det \gamma(i)}} \exp\left[\frac{i}{2\vartheta(s)}\left(\frac{d\gamma^{(i)}}{ds}\gamma^{(i)^{-1}}\vec{\varkappa}, \vec{\varkappa}\right)\right], \qquad i = \xi, \eta, \tag{2.9}$$

where for the columns of the matrix $\gamma^{(i)}$ we adopt two linearly-independent solutions of the system of Euler equations for the functional $\mathfrak{J}$ (see §1) in its approximate form in the vicinity of the principal ray (see [2]):

$$\frac{d^2\vec{\gamma}}{ds^2} - \frac{\vartheta'(s)}{\vartheta(s)} \cdot \frac{d\vec{\gamma}}{ds} + \mathcal{K}(s)\vec{\gamma} = 0, \qquad \vec{\gamma}(s) = (\gamma_\xi(s), \gamma_\eta(s)). \tag{2.10}$$

The condition (1.2) of concentration of the field in the vicinity of the ray imposes the following restriction on the matrix $\gamma^{(i)}$:

$$\mathfrak{Im}\left(\frac{d\gamma^{(i)}}{ds}\gamma^{(i)-1}\vec{\varkappa}, \vec{\varkappa}\right) > 0. \tag{2.11}$$

For resonators stable in the first approximation (see [1, 2]) we can formulate on the entire axis† four linearly-independent solutions of Eqs. (2.10) such as to satisfy the conditions

$$\left.\begin{aligned} \vec{\gamma}_\kappa(s+L) &= e^{i\alpha_\kappa}\vec{\gamma}_\kappa(s), \\ \vec{\gamma}_\kappa^*(s+L) &= e^{-i\alpha_\kappa}\vec{\gamma}_\kappa^*(s), \end{aligned}\right\} \quad \kappa = 1, 2, \tag{2.12}$$

where α_κ are the Floquet exponents (the asterisk denotes the complex conjugate), and

$$\left.\begin{aligned} &\left(\frac{d\vec{\gamma}_\kappa}{ds}, \vec{\gamma}_e\right) - \left(\vec{\gamma}_e \frac{d\vec{\gamma}_\kappa}{ds}\right) = i\vartheta(s)\delta_{\kappa e}, \quad \kappa = 1, 2, \ e = 1, 2, \\ &\left(\frac{\partial\vec{\gamma}_\kappa}{\partial s}, \vec{\gamma}_e^*\right) - \left(\vec{\gamma}_e, \frac{d\vec{\gamma}_\kappa^*}{ds}\right) = 0, \qquad \delta_{\kappa e} = \begin{cases} 1, & \kappa = e, \\ 0, & \kappa \neq e, \end{cases} \end{aligned}\right\} \tag{2.13}$$

[the scalar product is interpreted here as $(f_1, f_2) = f_{1\xi} \cdot \bar{f}_{2\xi} + f_{1\eta} \cdot \bar{f}_{2\eta}$].

Inequality (2.11) holds for the vectors $\vec{\gamma}_1$ and $\vec{\gamma}_2$ (for $\vec{\gamma}_1^*$ and $\vec{\gamma}_2^*$ the inequality is in the opposite sense). Inasmuch as the matrixes $\gamma(i)$, $i = \xi, \eta$, satisfy the same equation (2.10) and there are only two linearly-independent vectors corresponding to the concentration condition (1.2), we shall assume that

†The solutions are sought in each arm of the resonator in the local coordinate system (s_j, ξ_j, η_j) and are matched at the mirrors according to the reflection laws of geometrical optics (see [2]).

$$\left.\begin{aligned} &\gamma^{(1)} = \gamma^{(2)} = \gamma, \qquad \gamma = \| \vec{\gamma}_1, \vec{\gamma}_2 \|, \\ &\varphi^{\circ} = \frac{1}{\sqrt{\det \gamma(s)}} \exp\left[\frac{i}{2\vartheta(s)}\left(\frac{d\gamma}{ds}\gamma^{-1}\vec{\varkappa}, \vec{\varkappa}\right)\right], \\ &\varphi_i^{\circ} = C_i \varphi^{\circ}, \quad i = \xi, \eta; \quad C_i = \text{const}. \end{aligned}\right\} \tag{2.14}$$

Equations (2.2), (2.7), and (2.14) describe the zeroth approximation for the solutions of (1.1) concentrated in the vicinity of the ray.

§ 3. Zeroth-Approximation Eigenfrequencies

In the specification of the coordinate system (s_j, ξ_j, η_j) in each arm of the resonator there is an arbitrariness contingent upon the reference origin or the length s_j and angle of rotation θ_{0j}.

We set

$$d_{j-1} \leqslant s_j \leqslant d_j, \quad j = 1, \ldots, N, \tag{3.1}$$

where $s_j = d_{j-1}$, $\xi_j = \eta_j = 0$ is the point of reflection from the $(j-1)$st mirror and $s_j = d_j$, $\xi_j = \eta_j = 0$ is the point of incidence on the j-th mirror.

The linearly-independent solutions $E^{\circ}_{\xi_j}$ and $E^{\circ}_{\eta_j}$ for the j-th arm are written in the form

$$\vec{E}^{\circ}_j = \vec{e}_j \sqrt[4]{\frac{\mu}{\varepsilon}} \sqrt{\frac{\det \gamma(d_{j-1})}{\det \gamma(s_j)}} \exp\left[\frac{i}{2\vartheta}\left(\frac{d\gamma(s_j)}{ds_j}\gamma^{-1}(s_j)\vec{\varkappa}_j, \vec{\varkappa}_j\right) + i\omega \int_{d_{j-1}}^{s_j} \frac{ds}{\vartheta(s)}\right], \tag{3.2}$$

where

$$\vec{E}^{\circ}_j = E^{\circ}_{\xi_j}\vec{e}_{\xi_j} + E^{\circ}_{\eta_j}\vec{e}_{\eta_j},$$

$$\vec{C}_j = C_{\xi_j}\vec{e}_{\xi_j} + C_{\eta_j}\vec{e}_{\eta_j}.$$

For oblique incidence of a wave on the surface of the mirror the reflection coefficient depends on the polarization of the incident wave. The ideal reflection conditions yield

$$R_{\parallel} = 1, \qquad R_{\perp} = -1, \tag{3.3}$$

where $R_{\parallel} = \dfrac{E_{\parallel\,\mathrm{refl}}}{E_{\parallel\,\mathrm{inc}}}$, the vector $\vec{E}_{\parallel}$ lies in the plane of reflection, $R_{\perp} = \dfrac{E_{\perp\,\mathrm{refl}}}{E_{\perp\,\mathrm{inc}}}$, and the vector $\vec{E}_{\perp}$ is perpendicular to the plane of reflection.

The vectors $\vec{E}^{\circ}_{\xi_j}$ and $\vec{E}^{\circ}_{\eta_j}$ intersect the plane of reflection at a certain angle θ_j, which depends on the properties of the resonator and the initial angle θ_{0j}.

We define θ_{0j+1} so that

$$\left.\vec{E}^{\circ}_{\xi_{j+1}}\right|_{s_j} = \left.\vec{E}_{\parallel\,\mathrm{refl}}\right|_{s_j}, \tag{3.4}$$

$$\left.\vec{E}^{\circ}_{\eta_{j+1}}\right|_{s_j} = \left.\vec{E}_{\perp\,\mathrm{refl}}\right|_{s_j}.$$

Then the solutions (3.2) of the Maxwell equations in the j-th and (j + 1)st arms of the resonator are matched at the j-th mirror by the following reflection matrices:

$$\vec{E}^{\,0}_{j+1}\Big|_{s_j} = g\,u(\theta_j)\,\vec{E}^{\,0}_j\Big|_{s_j},$$
$$g=\begin{pmatrix}1, & 0\\ 0, & -1\end{pmatrix},\quad u(\theta_j)=\begin{pmatrix}\cos\theta_j, & -\sin\theta_j\\ \sin\theta_j, & \cos\theta_j\end{pmatrix}. \tag{3.5}$$

We relate the vectors $\vec{\gamma}_{ji}(d_j)$ and $\vec{\gamma}_{(j+1)i}(d_j)$, $i=1,2,\ j=1,\ldots,N$, according to the laws of geometrical optics (see [2]) and reduce conditions (3.5) to the form

$$\vec{C}_{j+1}=g\,u(\theta_j)\,\vec{C}_j\sqrt{\frac{\det\gamma_j(d_{j-1})}{\det\gamma_j(d_j)}}\exp\left(i\omega\int_{d_{j-1}}^{d_j}\frac{ds}{\vartheta(s)}\right). \tag{3.6}$$

After traversal of the closed cycle we obtain

$$\vec{C}_{N+1}=G\,\vec{C}_1\sqrt{\frac{\det\gamma(0)}{\det\gamma(L)}}\exp\left(i\omega\int_0^L\frac{ds}{\vartheta(s)}\right), \tag{3.7}$$

where

$$G=g\cdot u(\theta N)\cdot g\cdot u(\theta_{N-1})\cdots g\cdot u(\theta_1).$$

The matrix G is orthogonal (being either a rotation matrix for even N or the product of a rotation matrix and a specular-reflection matrix for odd N) and has two eigenvalues $e^{i\theta_p}$, $p=1,2$ ($\theta_2=-\theta_1$ for even N, $\theta_2=-\theta_1+\pi$ for odd N), which are independent of the choice of θ_{0j} (given a different choice, the new matrix G' will be similar to the old one).

The closure conditions (1.4) lead to the following expression for the eigenfrequencies of the resonator:

$$\omega_p=\frac{2\pi q+\frac{1}{2}(\alpha_1+\alpha_2)-\theta_p}{\int_0^L\frac{1}{\vartheta(s)}ds},\quad p=1,2,\quad q\gg 1,\ \text{integer} \tag{3.8}$$

[for the eigenfunctions (3.2) $\vec{C}^{\,0}_j=\vec{C}_p$, where $\vec{C}_p$ is an eigenvector of G].

The problem of the eigenmodes in a three-dimensional multimirror resonator has already been solved. It was assumed, as in the case of a homogeneous medium and a two-mirror resonator, that there are two wave modes (TE and TM), which can be described by the scalar equation $\left(\Delta+\frac{\omega^2}{c^2}\right)u=0$ and boundary conditions $u\big|_{s_j}=0$ and $\frac{\partial u}{\partial n}\big|_{s_i}=0$. Under these assumptions the formulas obtained for the eigenfrequencies (see [2, 3]) coincide with (3.8) for a homogeneous medium, provided the resonator axis is in one plane.

In the present study we did not impose any preliminary assumptions on the polarization of the eigenmodes. We deduced from the closure that two linearly-polarized modes propagate in the resonator with different frequencies ω_1 and ω_2, those modes comprising the analogs of the TE and TM modes for an inhomogeneous medium.

§ 4. Derivation of Equations for Any Approximation

We formulate the subsequent approximations of the solutions of the system (1.1) in the form

$$\vec{E} = \frac{1}{\sqrt{\varepsilon}} \sum_{p=1}^{2} \sum_{\kappa=0}^{\infty} \vec{\mathcal{E}}_p^{\kappa} \omega_p^{-\frac{\kappa}{2}} \exp\left(i\omega_p \int_{s_0}^{s} \frac{ds}{\vartheta}\right),$$
$$\vec{H} = \frac{1}{\sqrt{\mu}} \sum_{p=1}^{2} \sum_{\kappa=0}^{\infty} \vec{\mathcal{H}}_p^{\kappa} \omega_p^{-\frac{\kappa}{2}} \exp\left(i\omega_p \int_{s_0}^{s} \frac{ds}{\vartheta}\right) ds, \tag{4.1}$$

where ω_p is determined by Eq. (3.8) and $\vec{\mathcal{E}}_p^{\,0}$ and $\vec{\mathcal{H}}_p^{\,0}$ are the solutions of Eqs. (2.7) and (2.8) (all the relations for $\vec{\mathcal{E}}^{\kappa}$ and $\vec{\mathcal{H}}^{\kappa}$ from § 2 are fulfilled for $\vec{\mathcal{E}}_p^{\kappa}$ and $\vec{\mathcal{H}}_p^{\kappa}$, hence we can use them furnished with the polarization index).

The frequency ω in the equations for the p-th polarization is assumed to be equal to

$$\omega = \omega_p + \frac{\delta_p^1}{\omega^{1/2}} + \frac{\delta_p^1}{\omega} + \ldots \quad . \tag{4.2}$$

We introduce the following notation: $P_{p,q\tau}^{\kappa}$ is a polynomial of degree κ in x_p and y_p with coefficients depending on s and is even for even κ and odd for odd κ, with $q = \xi, \eta$ and $\tau = \xi, \eta$; $\mathcal{P}_p^{\kappa}$ is a second-order matrix composed of such polynomials; $\tilde{\mathcal{P}}_p^{\kappa}$ is the analogous fourth-order matrix; $\overset{\circ}{V}{}_p^{\kappa}$ is a solution of the homogeneous system $A\overset{\circ}{V}{}_p^{\kappa} = 0$;

$$\overset{\circ}{V}{}_p^{\kappa} = \begin{pmatrix} \overset{\circ}{\mathcal{E}}{}_{p\xi}^{\kappa} \\ \overset{\circ}{\mathcal{E}}{}_{p\eta}^{\kappa} \\ -\overset{\circ}{\mathcal{E}}{}_{p\eta}^{\kappa} \\ -\overset{\circ}{\mathcal{E}}{}_{p\xi}^{\kappa} \end{pmatrix}$$

$$\varphi_{pi}^{\kappa} \equiv \frac{1}{\vartheta(s)} \overset{\circ}{\mathcal{E}}{}_{pi}^{\kappa}, \; i = \xi, \eta; \; \vec{\varphi}_p^{\kappa} = (\varphi_{p\xi}^{\kappa}, \varphi_{p\eta}^{\kappa}).$$

Substitution of the series (4.1) into the Maxwell equations yields the following system of equations for $\overset{\circ}{V}{}_p^{3}$:

$$A V_p^3 = B V_p^1 + \tilde{\mathcal{P}}_p^3 V_p^0, \tag{4.3}$$

where the matrices B and A are defined as in (2.6).

We recall that $\det A = 0$, so that the system (4.3) is solvable only if the right-hand side is orthogonal to all solutions of the conjugate homogeneous system. We shall require this to be so, inferring then that the functions φ_{pi}^1 must satisfy the parabolic equation with finite right-hand side. We shall demonstrate below that the solution of the equation $L\varphi = p^{\kappa}\varphi^{\circ}$ has the form $\varphi = q^{\kappa}\varphi^{\circ}$, where q^{κ} is a polynomial of the type p^{κ}.

Having satisfied Eqs. (1.1) with ever-increasing accuracy and taken account of the form of the solution in each preceding step, we obtain the following system by induction for any approximation:

$$AV_p^{\kappa+2} = BV_p^{\kappa} + \tilde{\mathcal{P}}_p^{3\kappa} V_p^{0}, \tag{4.4}$$

[A and B have the same form as in (2.6)].

In the case when the right-hand side satisfies the solvability requirements the solution of the system (4.4) has the form

$$\overset{K+2}{V_p} = \overset{0\ K+2}{V_p} + \overset{1\ K+2}{V_p}, \tag{4.5}$$

where $\overset{0\ K+2}{V_p}$ is the general solution of the homogeneous system (4.4) and $\overset{1\ K+2}{V_p}$ is the particular solution of the inhomogeneous system. It is readily verified that

$$\overset{1\ K+2}{V_p} = \overset{\sim\ 3K+2}{\mathcal{P}_p} \overset{0}{V_p}.$$

The solvability conditions for the system (4.4) reduce to parabolic equations with finite right-hand side for the functions $\overset{K}{\varphi}_{pi}$, or

$$L_p \overset{K}{\vec{\varphi}_p} = \left(\overset{3K}{\mathcal{P}_p} - \frac{\overset{K}{\delta_p}}{\vartheta^2(s)} I\right) \vec{C}_p \overset{0}{\varphi_p}, \tag{4.6}$$

where

$$L_p = \frac{\partial^2}{\partial x_p^2} + \frac{\partial^2}{\partial y_p^2} + \frac{2i}{\vartheta}\frac{\partial}{\partial s} - \frac{1}{\vartheta^2}(\mathcal{K}\vec{\varkappa}_p, \vec{\varkappa}_p),$$

$$\vec{\varkappa}_p = (x_p, y_p),$$

$$x_p = \sqrt{\omega_p}\,\xi, \quad y_p = \sqrt{\omega_p}\,\eta,$$

$$p = 1, 2.$$

Here $I = \begin{pmatrix} 1, & 0 \\ 0, & 1 \end{pmatrix}$, and $\overset{3K}{\mathcal{P}_p}$ depends on $\overset{n}{\delta_p}$, $n \leq K-1$.

The solution of Eqs. (4.6) can be formulated by means of the "creation" and "destruction" operators (see [1-5])

$$\begin{aligned} \Lambda_{pe} &= \frac{1}{i}(\vec{\gamma}_e, \nabla_p) - \frac{1}{\vartheta}\left(\frac{d\vec{\gamma}_e}{ds}, \vec{x}_p\right), \\ \Lambda_{p\bar{e}} &= \frac{1}{i}(\vec{\gamma}_e^{\,*}, \nabla_p) - \frac{1}{\vartheta}\left(\frac{d\vec{\gamma}_e^{\,*}}{ds}, \vec{x}_p\right), \end{aligned} \tag{4.7}$$

where $\vec{x}_p = (x_p, y_p)$, $\nabla_p = \left(\frac{\partial}{\partial x_p}, \frac{\partial}{\partial y_p}\right)$, and $\vec{\gamma}_e, \vec{\gamma}_e^{\,*}$ are complex-conjugate solutions of the Euler equation (2.10) satisfying conditions (2.12) and (2.13). The functions

$$U_{pmn} = (\Lambda_{p\bar{1}})^m (\Lambda_{p\bar{2}})^n \overset{0}{\varphi_p}, \quad m = 1, 2, \ldots, n = 1, 2, \ldots \tag{4.8}$$

are solutions of the homogeneous equation

$$L_p U_{pmn} = 0 \tag{4.9}$$

and have the form

$$U_{pmn} = \overset{m+n}{R_p} \overset{0}{\varphi_p},$$

where $\overset{m+n}{R_p}$ is a polynomial of degree $m+n$ in x_p, y_p with coefficients depending on s and is even for even $m+n$ and odd for odd $m+n$.

The polynomials R_p^{m+n} are linearly independent, and there are as many of them as there are monomials of the form x_p^{κ}, y_p^{e}, $\kappa+e=m+n$, so that any polynomial in the variables x_p, y_p is a linear combination of R_p^{m+n} (see [1]).

We represent each element of the matrix $\mathcal{P}_p^{3\kappa}$ in Eqs. (4.6) in the form

$$P_{p,\kappa,\tau}^{3\kappa} = \sum_{m+n\leq 3\kappa} f_{pmn}^{(\kappa,q,\tau)}(s)\, R_p^{m+n}, \tag{4.10}$$

where

$$f_{pmn}^{(\kappa,q,\tau)}(s) = \frac{1}{2\pi m!\, n!} \int_{-\infty}^{+\infty}\int_{-\infty}^{+\infty} P_{p,q\tau}^{3\kappa} \overset{\circ}{\varphi}_p \overline{u_{pmn}}\, dx_p\, dy_p$$

(see [5]).

In the case when $P_{p,q\tau}^{3\kappa}$ is equal to the sum of odd powers of x_p, y_p, we have

$$f_{pmn}^{(\kappa,q,\tau)} = 0, \quad \text{for} \quad m+n = 2e. \tag{4.11}$$

But when $P_{p,q,\tau}^{3\kappa}$ is even, then $f_{p\,mn}^{(\kappa,q,\tau)} = 0$ for $m+n = 2e+1$, $e = 0, 1, \ldots.$

We seek the solution of the system (4.6) in the form

$$\vec{\varphi}_p^{\kappa} = \sum_{m+n\leq 3\kappa} \vec{a}_{pmn}\, u_{pmn}, \tag{4.12}$$

where $\vec{a}_{pmn} = (a_{pmn\xi}, a_{pmn\eta})$, and a_{pmni}, $i = \xi, \eta$ are arbitrary functions of s.

From Eqs. (4.6) we obtain in each arm of the resonator

$$\begin{aligned} i\frac{d\vec{a}_{pmn}^{\,j}}{ds_j} &= \vartheta(s_j)\Phi_{pmn}^{(\kappa)}(s_j)\vec{c}_p, \quad m\neq 0,\ n\neq 0 \\ i\frac{d\vec{a}_{p00}^{\,j}}{ds_j} &= \left(\vartheta(s_j)\Phi_{p00}^{(\kappa)}(s_j) - \frac{\delta_p^{\kappa}}{\vartheta(s_j)} I\right)\vec{c}_p, \\ \Phi_{pmn} &= \left\| f_{p,mm}^{(\kappa,q,\tau)} \right\|. \end{aligned} \tag{4.13}$$

The right-hand side of the system (4.13) represents a complex vector function, which, when taken around the entire closed axis L, acquires an increment of the argument:

$$\vec{\mathcal{F}}(s+L) = e^{-i(m\alpha_1 + n\alpha_2 - \theta_p)}\, \vec{\mathcal{F}}(s).$$

§ 5. Boundary-Value Problems for $\vec{a}_{pmn}^{\,j}(s_j)$; Corrections to the Eigenfrequencies

At each mirror we introduce the local coordinates $(\zeta_{1j}, \zeta_{2j}, \zeta_{3j})$, where ζ_{3j} is the inward normal to the mirror surface and ζ_{2j}, ζ_{1j} are curvilinear orthogonal coordinates to the mirror surface; at the point of incidence the unit vector $\vec{e}_{\zeta_{1j}}$ lies in the plane of incidence, and $\vec{e}_{\zeta_{2j}}$ is perpendicular to it; the equation for the surface has the form $\zeta_{3j} = 0$.

In the local coordinates the boundary conditions (1.3) are equivalent to the equations

$$E_{(s_{ij})}\Big|_{s_{3j}=0} = 0, \quad i=1,2, \quad j=1,\dots,N, \tag{5.1}$$

where

$$E_{(s_{ij})} = E_{(s_{ij})\mathrm{inc}} + E_{(s_{ij})\mathrm{refl}}, \tag{5.2}$$

and $E_{(q_i)}$ are the covariant components of the vector $\vec{E}$ along the corresponding axes.

We now express $E_{(s_{ij})\,\mathrm{inc}}$ in terms of $E_{(q_j)}$, $q = s, \xi, \eta$, and $E_{(s_{ij})\mathrm{refl}}$ in terms of $E_{(q_{j+1})}$ according to the well-known formulas for the transformation of the components of a vector from one curvilinear coordinate system to another, and we decompose $\frac{\partial q_j}{\partial s_{ij}}$ and $\frac{\partial q_{j+1}}{\partial s_{ij}}$, $i=1,2,3$, $q=\xi,\eta,s$, in the transformation formulas into formal series in the vicinity of the point of reflection $s_{1j} = s_{2j} = s_{3j} = 0$, assuming for each polarization that $z_{p_{ij}} = \sqrt{\omega_p}\, s_{ij}$. We require fulfillment of the boundary conditions (5.1) up to $\omega_p^{-\frac{k}{2}}$ (in the k-th approximation), replacing all the polynomials involved in the boundary conditions from segments of the series by linear combinations of R_p^{m+n}.

Then from Eqs. (5.1) we obtain for the functions $\vec{a}_{pmn}$

$$\begin{gathered} \vec{a}^{\,j+1}_{pmn}(d_j) = G(\theta_j)\,\vec{a}^{\,j}_{pmn}(d_j) + q^{j}_{mn}, \\ q^{j}_{mn} = \begin{cases} q^{j}_{mn}, & m+n \le k, \\ 0, & m+n > k \end{cases} \end{gathered} \tag{5.3}$$

(the numbers q^{j}_{mn} depend on the geometry of the j-th mirror).

The closure conditions (1.4) lead to the requirement

$$\begin{gathered} \vec{a}^{\,j+N}_{pmn}(s_j)\, e^{-i(m\alpha_1 + n\alpha_2 - \theta_p)} = \vec{a}^{\,j}_{pmn}(s_j), \\ \left(\vec{a}^{\,j+N}_{pmn}(s_j) = \vec{a}^{\,j}_{pmn}(s_j + L)\right), \end{gathered} \tag{5.4}$$

where G is the matrix from (3.7).

The solutions of Eqs. (4.13) have the form

$$\vec{a}^{\,j}_{pmn}(s_j) = -i\int_{d_{j-1}}^{s_j} \Phi^{(k)}_{pmn}(s_j)\,\vec{c}_p\,\vartheta(s_j)\,ds + \vec{a}_{oj}, \quad m \ne 0, \quad n \ne 0,$$

and

$$\vec{a}^{\,j}_{poo}(s_j) = -i\int_{d_{j-1}}^{s_j} \left(\vartheta(s_j)\,\Phi^{(k)}_{poo}(s_j) - \frac{\delta_p^k}{\vartheta(s_j)}\,I\right)\vec{c}_p\,ds + \vec{a}_{oj}, \tag{5.5}$$

where $\vec{a}_{oj} = (a^{j}_{o\xi}, a^{j}_{o\eta})$, and $a^{j}_{o\xi}$ and $a^{j}_{c\eta}$ are arbitrary constants.

The relationship (5.3) between the solutions in adjacent arms of the resonator is satisfied by the choice of $a^j_{o\xi}$, and $a^j_{o\eta}$.

In the sums (4.12) $m+n$ is made either even or odd, but Eq. (4.11) and the boundary conditions (5.4) imply the following:

$\vec{a}^j_{pmn} = (0,0)$, $m+n=2e$, for odd approximations;

$\vec{a}^j_{pmn} = (0,0)$, $m+n=2e+1$, for even approximations.

The self-adjoint problems (4.13) and (5.4) are solvable if the corresponding homogeneous problems do not have solutions.

For $m\alpha_1 + n\alpha_2 = 2\pi\tau$ (where τ is an integer) the homogeneous system

$$i\frac{d\vec{a}^j_{pmn}}{ds} = 0 \tag{5.6}$$

has a nontrivial solution satisfying condition (5.4):

$$a^{j+N}_{pmn} = Ga^j_{pmn}, \quad \vec{a}^j_{pmn} = \vec{c}_p, \tag{5.7}$$

where $\vec{c}_p$ is an eigenvector of the matrix G [cf. (3.7)].

We require solvability of problems (4.13) and (5.4) with an appropriate specification of δ^κ_p, i.e., bearing (4.11) in mind, we obtain

$$\delta^\kappa_p = \begin{cases} 0, & \kappa = 2e+1, \\ \dfrac{(\Phi^{(\kappa)}_{p00}\vartheta\vec{c}_p, \vec{c}_p)}{(\frac{1}{\vartheta}\vec{c}_p, \vec{c}_p)}, & \kappa = 2e, \end{cases} \tag{5.8}$$

where the scalar product of two vectors $\vec{f}(s)$ and $\vec{g}(s)$ has the standard definition

$$(\vec{f},\vec{g}) = \int_s^{s+L} \sum_{i=\xi,n} f_i(s)\overline{g_i(s)}ds.$$

Now the boundary-value problems (4.13) and (5.4) are solvable for any approximation if the numbers α_1, α_2 and 2π are linearly independent over the ring of integers ($m\alpha_1 + n\alpha_2 + \tau\cdot 2\pi = 0$, where m, n, and τ are integers subject to the constraint $m = n = \tau = 0$); Eqs. (5.8) describe the corrections to the eigenvalues (3.8).

§ 6. Higher Harmonics

The eigenfunctions and eigenfrequencies formulated in §§ 3-5 determine two fundamental harmonics in the resonator (longitudinal modes) that differ only in their polarizations. The higher harmonics (or transverse modes of the resonator) are determined in the zeroth approximation by Eqs. (4.8). Writing these formulas in the local coordinate system for each arm of the resonator, as a result of their "reflection and "closure" (as in § 3, with regard for the properties of the operators $\Lambda_{p\kappa}$; see also [2]), we write the following relation for the eigenfrequencies:

$$\omega_{p,qmn} = \frac{2\pi q + (m+\frac{1}{2})\alpha_1 + (n+\frac{1}{2})\alpha_2 - \theta_p}{\int_{s_0}^{s_0+L} \frac{ds}{\vartheta(s)}} \tag{6.1}$$

(where P is the polarization index, $p = 1, 2$, $q >> 1$ is an integer, and $m, n = 1,2,3,\ldots$), which is the case of a homogeneous medium and plane polygon (resonator axis) coincides with the result obtained earlier for TE and TM modes (see [2, 3]).

The approximations of the eigenfunctions of the system of Maxwell equations can be formulated with any accuracy in the form of series (4.1), which are rewritten for $\omega_{p,qmn}$, respectively:

$$\vec{E}_{p,qmn} = \frac{1}{\sqrt{\varepsilon}} \sum_{K=0}^{\infty} \vec{\mathcal{E}}^{K}_{p,qmn} \, \omega_{p,qmn}^{-\frac{K}{2}},$$
$$\vec{H}_{p,qmn} = \frac{1}{\sqrt{\mu}} \sum_{K=0}^{\infty} \vec{\mathcal{H}}^{K}_{p,qmn} \, \omega_{p,qmn}^{-\frac{K}{2}}, \tag{6.2}$$

where the zeroth approximation $(K=0)$ is determined by means of (4.8):

$$\omega = \omega_{p,qmn} + \frac{\delta^{1}_{p,qmn}}{\sqrt{\omega_{p,qmn}}} + \ldots . \tag{6.3}$$

Following the same procedure as before (see § 4), we obtain inhomogeneous parabolic equations

$$L_p \vec{\varphi}^{K}_{p,mn} = \left(\mathcal{P}_p^{m+n+3K} - \frac{\delta^{K}_{p,qmn}}{\vartheta^2(s)} \right) \vec{C}_p \varphi_p^0, \tag{6.4}$$

where

$$\vec{\varphi}^{K}_{p,mn} = \left(\varphi^{K}_{p\xi,mn}, \varphi^{K}_{p\eta,mn} \right),$$

$$\varphi^{K}_{pi,mn} \equiv \frac{1}{\sqrt{\vartheta(s)}} \overset{0}{\mathcal{E}}{}^{K}_{pi,mn}, \quad i = \xi, \eta.$$

We replace the polynomials on the right-hand sides of (6.4) by linear combinations of R_p^{s+t}:

$$P_{pi\tau}^{m+n+3K} = \sum_{s+t \leq m+n+3K} f_{p\,st}^{(K,m,n,i,\tau)} R_p^{s+t}, \tag{6.5}$$

where not all the coefficients $f_{p,st}$ in the sums (6.5) have nonzero values.

Consider the first approximation, $K = 1$.

Equations (6.4) initially had the form

$$L_p \vec{\varphi}^{1}_{p,mn} = D \vec{C}_p (\Lambda_{p\bar{1}})^m (\Lambda_{p\bar{2}})^n \varphi_p^0, \tag{6.6}$$

where D is a matrix whose elements are linear combinations of

$$P^1, \quad P^2 \frac{\partial}{\partial x}, \quad P^2 \frac{\partial}{\partial y}, \quad P^1 \frac{\partial}{\partial s}, \quad P^3,$$

and P^K are different polynomials of degree K, as in § 4. Each element of D may be represented in the form

$$\sum_{i=1,2,\bar{1},\bar{2}} a_i \Lambda_{pi} + \sum_{\substack{i=1,2,\bar{1},\bar{2}\\ \kappa=1,2,\bar{1},\bar{2}\\ \tau=1,2,\bar{1},\bar{2}}} \Lambda_{pi}\Lambda_{p\kappa}\Lambda_{p\tau} + p' L_p. \tag{6.7}$$

Utilizing the properties of the operators Λ_{pi} and the orthogonality of the functions $\mathcal{U}_{pmn}$, we can readily show that $f_{p,st} \neq 0$ only for the following combinations of the indices s and t:

$$\frac{s}{t} = \frac{m\pm 1}{n}, \frac{m\pm 3}{n}, \frac{m\pm 2}{n\pm 1}, \frac{m\pm 1}{n\pm 2}, \frac{m}{n\pm 3}, \frac{m}{n\pm 1} \tag{6.8}$$

Analogous "selection rules" prevail in the equations of the next higher approximation, where it is easily verified that

$$f_{p,st} \neq 0, \quad \text{for} \quad s=m,\ t=n \tag{6.9}$$

only for even approximations.

As in the case of the fundamental harmonic, we seek the solution of Eqs. (6.4) in the form

$$\vec{\varphi}^{\,\kappa}_{p,mn} = \sum_{s,t} \vec{a}_{pst}\, \mathcal{U}_{pst}. \tag{6.10}$$

Equations (6.4) and the closure conditions (1.4) lead in each arm to problems analogous to (4.13) and (5.4), their solutions $\vec{a}^{\,j}_{p,st}(d_j)$ and $\vec{a}^{\,j+1}_{p,st}(d_j)$ being related at each j-th mirror by boundary conditions of the type (5.3) that are satisfied by proper choice of the arbitrary constants of integration (see § 5). The summation indices s and t in (6.10) obey the same selection rules as in (6.6) (the closure conditions imply that $\vec{a}_{pst} = 0$ for all other combinations of s and t).

The corrections to the eigenvalues (6.1), as before, are obtained from the requirement of solvability of the boundary-value problems for $\vec{a}^{\,i}_{pmn}(s)$.

In conclusion the author would like to thank V. M. Babich for supervising the present study.

LITERATURE CITED

1. Babich, V. M., Eigenfunctions concentrated in a neighborhood of a closed geodesic, Seminars in Mathematics, Vol. 9: Mathematical Problems in Wave Propagation Theory, Consultants Bureau, New York (1970), pp. 7-26.
2. Popov, M. M., Eigenmodes of multimirror resonators, Vest. Leningrad. Univ. (in press).
3. Popov, M. M., Asymptotic Behavior of the Eigenfunctions of the Helmholtz Equation and Its Application to the Theory of Multimirror Resonators, Author's abstract of dissertation for the degree Candidate of Physicomathematical Sciences, Leningrad (1969).
4. Lazutkin, V. F., Spectral degeneracy and "small denominators" in the asymptotic representation of eigenfunctions of the "boundary ball" type, Vest. Leningrad. Univ., No. 7 (1969).
5. Kirpichnikova, N. Ya., On the formulation of solutions concentrated near rays of the elasticity equations for an inhomogeneous isotrotropic space, Seminars in Mathematics (in press).

HIGH APPROXIMATIONS FOR THE EIGENMODES OF A PLANE MULTIMIRROR RESONATOR

T. M. Popova

The present article is devoted to an investigation of the properties of a multimirror resonator. The object of the investigation is to formulate the eigenfunctions and eigenfrequencies for a plane resonator in any approximation by the parabolic equation method. The plan of the method represents an extension of the plan carried out in [1] for the double-mirror resonator problem.

Consider a multimirror resonator formed by M ideal mirrors with radii of curvature ρ_j, $j=1,2,\ldots,M$, situated at the vertices of an M-sided polygon with sides $AB=d_1,\ldots,DA=d_M$ (Fig. 1). Let the mirrors be mounted at the vertices with angles $2\delta_1,\ldots,2\delta_M$ such that their centers of curvature lie on the bisectors of the corresponding angles. Then the sides of the polygon are generated in sequence, one from another, by reflection in the mirrors according to the law of geometrical optics: The angle of incidence is equal to the angle of reflection.

In the j-th arm of the resonator we introduce a rectangular coordinate system x_j', z_j (Fig. 2).

The coordinate systems x_j', z_j and ν, t are related by the equations

$$z_j = d_j - \nu\cos\delta_j + t\sin\delta_j \qquad z_{j+1} = \nu\cos\delta_j + t\sin\delta_j$$

$$x_j' = -\nu\sin\delta_j - t\cos\delta_j \qquad x_{j+1}' = -\nu\sin\delta_j + t\cos\delta_j,$$

where $j=1,2,\ldots,M$.

We are seeking eigenmodes of the resonator that are concentrated near its axis and decay exponentially with distance from it. The problem is stated as follows: Find in the vicinity of the closed polygon a solution of the equation

$$(\Delta + k^2)U = 0 \quad \text{for} \quad k \to \infty \tag{1}$$

satisfying the condition

$$U|_{s_j} = 0, \qquad j = 1,2,\ldots,M \tag{2}$$

at the mirrors and decaying exponentially with distance from the polygon.

We shall assume that the mirror is described by the equation

$$\nu = \sum_{k=2}^{\infty} R_k t^k = \sum_{k=2}^{\infty} \frac{R_k}{N^k}\tau^k; \quad N=\sqrt{k}; \quad R_2 = \frac{1}{2\rho},$$

where $\tau = Nt$.

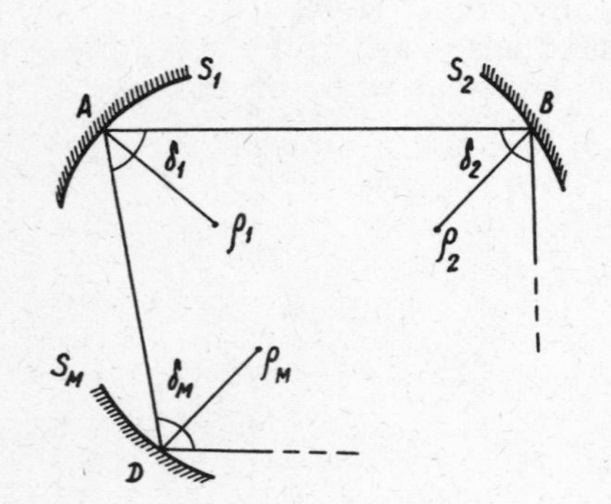

Fig. 1.

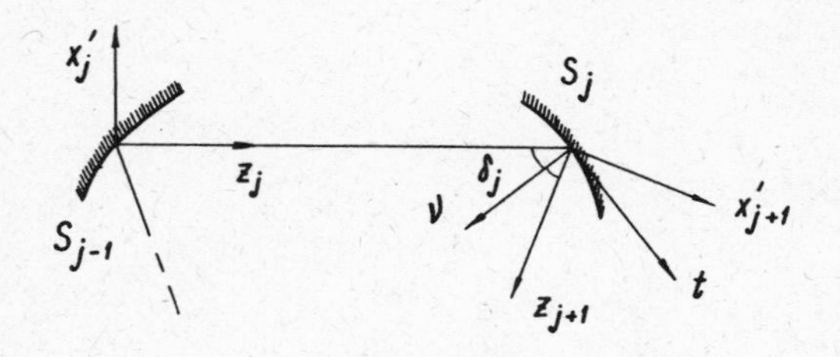

Fig. 2.

In accordance with the parabolic equation method, instead of x_j' we consider the variable $x_j = \sqrt{k}\, x_j' = N x_j'$. We seek the solution of the problem (1)-(2) in the form

$$U = \sum_{j=1}^{M} U^{(j)},$$

where each term refers to the corresponding arm of the resonator and is expressed by the formula

$$U^{(j)} = e^{i\Phi^{(j)}} \mathcal{D}_n (\Psi^{(j)}), \tag{3}$$

where $\mathcal{D}_n(\Psi^{(j)})$ is a parabolic cylindrical function satisfying the equation

$$\mathcal{D}_n'' + [\varkappa_n - (\Psi^{(j)})^2]\, \mathcal{D}_n = 0, \quad \varkappa_n = 2n+1, \quad n = 0, 1, 2, \ldots$$

and decaying exponentially as $|\Psi^{(j)}| \to \infty$.

We make the following assumptions with regard to the functions $\Psi^{(j)}$ and $\Phi^{(j)}$:

$$\Phi^{(j)}(z_j, x_j) = N^2 z_j + \sum_{p=2}^{\infty} \frac{\varphi_p^{(j)}(z_j, x_j)}{N^{p-2}},$$
$$\Psi^{(j)}(z_j, x_j) = \sum_{p=1}^{\infty} \frac{1}{N^{p-1}} \psi_p^{(j)}(z_j, x_j), \tag{4}$$

where the following notation is used:

$$\varphi_p^{(j)}(z, x) = \varphi_{p,p}^{(j)}(z) \cdot x^p + \cdots + \varphi_{p,0}^{(j)}(z),$$
$$\psi_p^{(j)}(z, x) = \psi_{p,p}^{(j)}(z) \cdot x^p + \cdots + \psi_{p,0}^{(j)}(z). \tag{5}$$

Consequently, in order to find the solution of the problem (1)-(2) it is neccessary to deremine the coefficients in the decomposition of the functions $\Phi^{(j)}$ and $\Psi^{(j)}$ in powers of $\frac{1}{N}$. Substituting the function (3) into Eq. (1) and setting the coefficients of like powers of $\frac{1}{N}$ equal to zero and then the same for like powers of x_j, we obtain, after [1], a recursive system of equations for determining the coefficients in the decompositions of $\Phi^{(j)}$ and $\Psi^{(j)}$.

The boundary conditions for the functions $\Phi^{(j)}$ and $\Psi^{(j)}$ arise from condition (2) and have the form

$$\Psi^{(j)}\big|_{S_j} = \Psi^{(j+1)}\big|_{S_j},$$
$$\Phi^{(j)}\big|_{S_j} = \Phi^{(j+1)}\big|_{S_j} + \text{const}. \tag{6}$$

At the mirror the coordinates x_j, z_j, and τ are related by the expressions

$$
\begin{aligned}
z_j &= d_j + \frac{\sin\delta_j}{N}\tau - \cos\delta_j \cdot \sum_{k=2}^{\infty} \frac{R_k^{(j)}}{N^k}\tau^k \equiv d_j + \sum_{k=1}^{\infty} \frac{A_k^{(j)}}{N^k}\tau^k, \\
x_j &= -\tau\cos\delta_j - \sin\delta_j \cdot \sum_{k=2}^{\infty} \frac{R_k^{(j)}}{N^{k-1}}\tau^k \equiv \sum_{k=0}^{\infty} B_{k+1}^{(j)} \frac{\tau^{k+1}}{N^k}, \\
z_{j+1} &= \cos\delta_j \cdot \sum_{k=2}^{\infty} \frac{R_k^{(j)}}{N^k}\tau^k + \frac{\tau}{N}\sin\delta_j \equiv \sum_{k=1}^{\infty} \frac{A_k^{(j+1)}}{N^k}\tau^k, \\
x_{j+1} &= \tau\cos\delta_j - \sin\delta_j \cdot \sum_{k=2}^{\infty} \frac{R_k^{(j)}}{N^{k-1}}\tau^k \equiv \sum_{k=0}^{\infty} B_{k+1}^{(j+1)} \frac{\tau^{k+1}}{N^k}.
\end{aligned}
\tag{7}
$$

We substitute the decompositions (7) into the boundary conditions (6) and decompose the resulting expressions in powers of $\frac{1}{N}$. Writing out the coefficients of like powers of $\frac{1}{N}$ on both sides of the boundary conditions (6) and separating terms of like order on τ, we obtain the boundary conditions that must be satisfied by $\varphi^{(j)}_{p+1,p+1}, \ldots, \varphi^{(j)}_{p+1,0}$, and $\psi^{(j)}_{p,p}, \ldots, \psi^{(j)}_{p,0}$ for $p = 1, 2, \ldots$.

We thus encounter the problem of determining the coefficients of the polynomials $\varphi_p^{(j)}$ and $\psi_p^{(j)}$ (5).

We now consider briefly the solutions of the problem (1)-(2) in the first approximation. In the decomposition (4) we reject terms of order $O(\frac{1}{N})$ and higher:

$$
\begin{aligned}
\Phi^{(j)}(z_j, x_j) &= N^2 z_j + \varphi_2^{(j)}(z_j, x_j), \\
\Psi^{(j)}(z_j, x_j) &= \psi_1^{(j)}(z_j, x_j),
\end{aligned}
$$

where

$$
\begin{aligned}
\varphi_2^{(j)}(z_j, x_j) &= \varphi_{2,2}^{(j)}(z_j)\cdot x_j^2 + \varphi_{2,0}^{(j)}(z_j), \\
\psi_1^{(j)}(z_j, x_j) &= \psi_{1,1}^{(j)}(z_j)\cdot x_j .
\end{aligned}
$$

In the first approximation for the coefficients $\Psi^{(j)}$ and $\Phi^{(j)}$ we obtain the system of equations

$$
\begin{gathered}
2i\frac{\partial \psi_1^{(j)}}{\partial z_j} + 2i\frac{\partial \varphi_2^{(j)}}{\partial x_j}\cdot\frac{\partial \psi_1^{(j)}}{\partial x_j} + \frac{\partial^2 \psi_1^{(j)}}{\partial x_j^2} = 0, \\
\cdot 2\frac{\partial \varphi_2^{(j)}}{2 z_j} + i\frac{\partial^2 \varphi_2^{(j)}}{\partial x_j^2} - 2\left(\frac{\partial \varphi_2^{(j)}}{\partial x_j}\right)^2 + 4\left(\frac{\partial \psi_1^{(j)}}{\partial x_j}\right)^2 (\psi_1^{(j)})^2 - 2\varkappa_n \left(\frac{\partial \psi_1^{(j)}}{\partial x_j}\right)^2 = 0, \\
j = 1, 2, \ldots, M.
\end{gathered}
$$

The solution of this system has the form

$$\varphi_{1,1}^{(j)} = \frac{1}{\theta_j(z_j)}, \quad \varphi_{2,2}^{(j)} = \frac{1}{2}\frac{\theta_j'}{\theta_j},$$

$$\varphi_{2,0}^{(j)} = -\left(n+\frac{1}{2}\right)\int_0^{z_j} \frac{dt}{\theta_j^2(t)} + i\cdot\frac{1}{2}\ln|\theta_j|,$$

where $\theta_j(z_j)=\sqrt{\frac{1}{\lambda_j}+\lambda_j(z_j-\varsigma_j)^2}$, λ_j and ς_j are arbitrary constants. From the boundary conditions (6) on the j-th mirror we infer the following equations for the functions $\theta_j(z_j)$ with retention of only the $\frac{1}{2\rho_j}\cdot\frac{\tau^2}{N^2}$ in the mirror equation:

$$\theta_{j+1}(0) = -\theta_j(d_j),$$

$$\frac{\theta_j'(d_j)}{\theta_j(d_j)} = \frac{\theta_{j+1}'(0)}{\theta_{j+1}(0)} + \frac{2}{\rho_j\cos\delta_j}. \tag{8}$$

In order for the solution of the problem in the form (3) to decay exponentially with distance from the resonator axis it is required that the function θ be real.

It can be shown that the boundary conditions (8) are fulfilled if certain parameters are linearly transformed by the ray-theory reflection matrix method [3]. We write the function θ_j in the form $\theta_j=\sqrt{\alpha_j z^2+2\beta_j z+\gamma_j}$, and from the coefficients α_j, β_j, and γ_j we form the matrix Ω_j according to the formula

$$\Omega_j = \begin{Vmatrix} \alpha_j & \beta_j \\ \beta_j & \gamma_j \end{Vmatrix}, \quad j=1,2,\ldots,M,M+1.$$

We introduce the matrix P_j, which is related to Ω_j by the formula

$$\Omega_j = (P_j P_j^T)^{-1}.$$

We see at once that P_j is determined from Ω_j up to an orthogonal matrix.

Let us construct $P_{j+1} = T_j P_j$, where

$$T_j = \begin{Vmatrix} -1 & d_j \\ -\frac{2}{\rho_j\cos\delta_j} & \frac{2d_j}{\rho_j\cos\delta_j}-1 \end{Vmatrix}$$

is the ray-theory reflection matrix at the j-th mirror; let us then set

$$\Omega_{j+1} = (P_{j+1}P_{j+1}^T)^{-1},$$

whereupon θ_j and θ_{j+1} satisfy the boundary conditions (8).

After one bypass of the resonator $P_{M+1} = \Gamma P_1$, where $\Gamma = T_M\cdots T_1$. In order to obtain the eigenfunction it is required that $\theta_{N+1}^2 = \theta_1^2$.*

*The choice of sign for the function θ is dictated by whether the number of mirrors in the resonator is even or odd.

The condition of periodicity for the function θ^2 can be written in the matrix form $\Gamma Q \Gamma^T = Q$, where $Q = P_1 P_1^T$ is a symmetric matrix. This equation is solvable in Q only if the resonator is stable in the first approximation in the sense delineated in [2, 3] from geometrical-optic considerations.

The solution of the problem (1)-(2) in the first approximation has the form

$$U^{(j)} = \frac{C^{(j)}}{\sqrt{|\theta_j|}} \mathcal{D}_n\left(\frac{x_j' \sqrt{K_{pn}}}{\theta_j(z_j)}\right) \cdot \exp i\left\{K_{pn}\left(z_j + \frac{1}{2}\frac{\theta_j'}{\theta_j} x_j^2\right) - \left(n+\frac{1}{2}\right)\int_0^{z_j} \frac{dz}{\theta_j^2(z)}\right\},$$

where $n = 0, 1, 2, \ldots$, and

$$1)\quad C_{j+1} = C_j \exp i\left\{\pi + K_{pn} d_j - \left(n+\frac{1}{2}\right)\int_0^{d_j} \frac{dz}{\theta_j^2(z)}\right\},$$

$$2)\quad K_{pn} = \frac{1}{L}\left\{\pi(2p - M + \delta) + \left(n+\frac{1}{2}\right)\sum_{j=1}^{M} \int_0^{d_j} \frac{dz}{\theta_j^2(z)}\right\},$$

where $L = \sum_{j=1}^{M} d_j$, $p \gg 1$, and

$$\delta = \begin{cases} 1 & \text{for } M \text{ and } n \text{ odd;} \\ 0 & \text{in all other cases.} \end{cases}$$

Next we formulate the subsequent approximations. We assume that for some arbitrary integer τ a solution of the problem (1)-(2) has been obtained up to terms of order $O\left(\frac{1}{N^{\tau-1}}\right)$ and find the solution of the problem up to terms of order $O\left(\frac{1}{N^{\tau}}\right)$.

Thus by the induction method we can formulate the solution in any approximation. The equations for $\Psi_{\tau+3}^{(j)}$ and $\varphi_{\tau+4}^{(j)}$ are particularly simple to exhibit if the following change of variables is instituted:

$$\tilde{z}_j = \int_0^{z_j} \frac{ds}{\theta^2(s)}, \quad \eta_j = \frac{1}{\theta(z_j)} x_j, \quad \alpha_j = \int_0^{d_j} \frac{ds}{\theta^2}.$$

In the new variables we have

$$\begin{aligned} \varphi_{\tau+4}^{(j)}(z_j, x_j) &= \tilde{\varphi}_{\tau+4,\tau+4}^{(j)} \cdot \eta_j^{\tau+4} + \cdots + \tilde{\varphi}_{\tau+4,0}^{(j)}, \\ \Psi_{\tau+3}^{(j)}(z_j, x_j) &= \tilde{\Psi}_{\tau+3,\tau+3}^{(j)} \cdot \eta_j^{\tau+3} + \cdots + \tilde{\Psi}_{\tau+3,0}^{(j)}. \end{aligned} \tag{9}$$

For the coefficients in (9), setting the corresponding term equal to zero in the decomposition in powers of $\frac{1}{N}$ of Eq. (1), we arrive at the equations

$$\frac{\partial \tilde{\varphi}_{\tau+4,\tau+4}^{(j)}}{\partial \tilde{z}_j} - (\tau+4)\tilde{\Psi}_{\tau+3,\tau+3}^{(j)} = 0,$$

$$\frac{\partial \tilde{\varphi}^{(j)}_{\tau+4,\tau+3}}{\partial \tilde{z}_j} - (\tau+3)\, \tilde{\psi}^{(j)}_{\tau+3,\tau+2} = 0,$$

$$\cdots\cdots\cdots\cdots\cdots\cdots$$

$$\frac{\partial \tilde{\varphi}^{(j)}_{\tau+4,0}}{\partial \tilde{z}_j} - i\, \tilde{\varphi}^{(j)}_{\tau+4,2} + \varkappa_n \tilde{\psi}^{(j)}_{\tau+3,1} = F,$$

$$\frac{\partial \tilde{\psi}^{(j)}_{\tau+3,\tau+3}}{\partial \tilde{z}_j} + (\tau+4)\, \tilde{\varphi}^{(j)}_{\tau+4,\tau+4} = 0,$$

$$\frac{\partial \tilde{\psi}^{(j)}_{\tau+3,\tau+2}}{\partial \tilde{z}_j} + (\tau+3)\, \tilde{\varphi}^{(j)}_{\tau+4,\tau+3} = F,$$

$$\cdots\cdots\cdots\cdots\cdots\cdots$$

$$\frac{\partial \tilde{\psi}^{(j)}_{\tau+3,0}}{\partial \tilde{z}_j} + \tilde{\varphi}^{(j)}_{\tau+4,1} - i\, \tilde{\psi}^{(j)}_{\tau+3,2} = F.$$

Here F depends on the functions already determined by the induction hypothesis.*

The boundary conditions for the coefficients of the polynomials (9) are obtained analogously from the boundary conditions (2):

$$\tilde{\psi}^{(j)}_{\tau+3,m-1}(\alpha_j) = \tilde{\psi}^{(j+1)}_{\tau+3,m-1}(0) + F,$$

$$\tilde{\varphi}^{(j)}_{\tau+4,m}(\alpha_j) \frac{1}{\theta_j^m(d_j)} - \tilde{\varphi}^{(j+1)}_{\tau+4,m}(0) \frac{(-1)^m}{\theta_{j+1}^m(0)} = F,$$

where $m = 1, \ldots, \tau+4$.

Thus we arrive at the following problem of determining the coefficients of the polynomials $\varphi^{(j)}_{\tau+4}$ and $\psi^{(j)}_{\tau+3}$. Find a solution, periodic on the polygon, of the following system of equations:

$$\begin{aligned} &\frac{\partial \tilde{\varphi}^{(j)}_{\tau+4,m}}{\partial \tilde{z}_j} - m\, \tilde{\psi}^{(j)}_{\tau+3,m-1} = F, \\ &\qquad\qquad\qquad\qquad\qquad m = 1, \ldots, \tau+4, \\ &\frac{\partial \tilde{\psi}^{(j)}_{\tau+3,m-1}}{\partial \tilde{z}_j} + m\, \tilde{\varphi}^{(j)}_{\tau+4,m} = F, \end{aligned} \tag{10}$$

satisfying the following conditions at the vertices of the polygon:

$$\begin{aligned} &\tilde{\varphi}^{(j)}_{\tau+4,m}(\alpha_j) = \tilde{\varphi}^{(j+1)}_{\tau+4,m}(0) + F, \\ &\qquad\qquad\qquad\qquad\qquad m = 1, \ldots, \tau+4. \\ &\tilde{\psi}^{(j)}_{\tau+3,m-1}(\alpha_j) = \tilde{\psi}^{(j+1)}_{\tau+4,m-1}(0) + F. \end{aligned} \tag{11}$$

We now examine the solvability aspect of the problem. Let $F = 0$ in the problem (10)-(11), i.e., let the problem be homogeneous. We seek, for example, the leading coefficients of the polynomials $\tilde{\varphi}^{(j)}_{\tau+4,\tau+4}$

---

*Note that here and everywhere below F differs and denotes combinations of functions already known.

and $\tilde{\Psi}^{(j)}_{\tau+3,\tau+3}$ on the j-th arm. It follows from Eqs. (10) that

$$\begin{aligned}\tilde{\varphi}^{(j)}_{\tau+4,\tau+4} &= a_j \cos(\tau+4)\tilde{z}_j + b_j \sin(\tau+4)\tilde{z}_j;\\ \tilde{\Psi}^{(j)}_{\tau+3,\tau+3} &= -a_j \sin(\tau+4)\tilde{z}_j + b_j \cos(\tau+4)\tilde{z}_j,\end{aligned} \tag{12}$$

where a_j and b_j are arbitrary constants.

We denote the arbitrary constants for $\tilde{\varphi}^{(j+1)}_{\tau+4,\tau+4}$ and $\tilde{\Psi}^{(j+1)}_{\tau+3,\tau+3}$ in the $(j+1)$ st arm by a_{j+1} and b_{j+1}. From Eqs. (11) we obtain

$$\begin{pmatrix} a_{j+1} \\ b_{j+1} \end{pmatrix} = \begin{pmatrix} \cos(\tau+4)\alpha_j & \sin(\tau+4)\alpha_j \\ -\sin(\tau+4)\alpha_j & \cos(\tau+4)\alpha_j \end{pmatrix} \begin{pmatrix} a_j \\ b_j \end{pmatrix} \equiv g^{(\tau+4)}_{\alpha_j} \begin{pmatrix} a_j \\ b_j \end{pmatrix},$$

i.e., in the transition from one arm of the resonator to the next the constants of integration are linearly transformed by means of the orthogonal matrix $g^{(\tau+4)}_{\alpha_j}$.

After traversal of the polygon we end up with the matrix

$$g^{(\tau+4)}_{\varphi} = g^{(\tau+4)}_{\alpha_M} \cdots g^{(\tau+4)}_{\alpha_1}; \qquad \varphi = \sum_{j=1}^{M} \alpha_j,$$

where M is the number of vertices of the polygon. Requiring fulfillment of the periodicity conditions, we obtain a system of equations for determining the constants of integration a_1, and b_1:

$$\| g^{(\tau+4)}_{\varphi} - E \| \begin{pmatrix} a_1 \\ b_1 \end{pmatrix} = 0.$$

This system will have only a trivial solution of

$$\det \| g^{(\tau+4)}_{\varphi} - E \| \neq 0.$$

The latter condition is equivalent to the assertion that

$$1 - \cos(\tau+4)\varphi \neq 0,$$

whence we arrive at the following restriction on φ:

$$(\tau+4)\varphi \neq 2\pi n, \qquad n = 0, 1, 2, \ldots. \tag{13}$$

If the requirement (13) is fulfilled, a solution satisfying conditions (11) at the vertices of the polygon and periodic on that polygon exists for the inhomogeneous system (10) and is unique.

Consequently, if φ is noncommensurate with 2π, the procedure outlined above makes it possible to formulate an approximation solution of the problem (1)-(2) with arbitrarily small error for $N = \sqrt{k} \to \infty$, i.e., to solve the problem with any accuracy. Otherwise these formulations are cut off in some step.

The author is deeply grateful to his academic sponsor V. M. Babich for help offered in the study, as well as to V. F. Lazutkin and M. M. Popov for a discussion of the article.

LITERATURE CITED

1. Lazutkin, V. F., Candidate's Dissertion, V. A. Steklov Mathematical Institute, Leningrad (1967).
2. Buldyrev, V. S., Asymptotic behavior of the eigenfunctions of the Helmoltz equation for plane convex domains, Vest. Leningrad. Univ., No. 22(4), p. 38 (1965).
3. Popov, M. M., Geometrical optics and the eigenfrequencies of annular resonators, Vest. Leningrad. Univ., No. 4(1), pp. 42-52 (1967).

ASYMPTOTIC BEHAVIOR OF EIGENFUNCTIONS OF THE HELMHOLTZ EQUATION CONCENTRATED NEAR A CLOSED GEODESIC

M. F. Pyshkina

In the present article we wish to examine the asymptotic behavior of eigenfunctions concentrated in the vicinity of a closed geodesic on an $(m+1)$-dimensional differentiable compact Riemann manifold. In the article we shall make extensive use of the mathematical apparatus and notation of [1], tacitly assuming prior acquaintance with that paper.

The eigenfunctions in our problem are determined as the nonzero solutions of the equation

$$(\Delta + \kappa^2)u = 0, \tag{1}$$

where Δ is the Laplace operator. If g_{ij} is a metric tensor in the coordinates y_i, then

$$\Delta u \equiv \frac{1}{\sqrt{g}} \frac{\partial}{\partial y_i}\left(\sqrt{g}\, g^{ij} \frac{\partial u}{\partial y_i}\right).$$

All of the ensuing formulations are conveniently realized in the same coordinates as in [1]: $(s, y_1, \ldots, y_m) = (y^0, y^1, \ldots, y^m)$.

In [1] an explicit expression was derived for the eigenfunctions and eigenvalues to a first approximation. More precisely, with the substitution into Eq. (1) of our resulting eigenfunction the error turns out to be of the order $O(\kappa^{1/2})$. In the first approximation the eigenvalues have the form

$$\kappa_{pq} = \frac{1}{L}\left[2\pi p + \sum_{j=1}^{m}\left(q_j + \frac{1}{2}\right)\alpha_j\right], \quad q = (q_1, \ldots, q_m), \tag{2}$$

where $p \gg 1$ is an integer and $q_j = O(1)$. Here α_j is the Floquet constant (cf. [1]). We have succeeded in finding an asymptotic series both for an eigenvalue whose principal term has the form (2) and for the corresponding eigenfunction.

We were able to carry out our formulations under the additional assumption of linear independence of the numbers $\pi, \alpha_1, \ldots, \alpha_m$ over the ring of integers (i.e., the equation

$$\pi d_0 + \sum_{j=1}^{m} \alpha_j d_j = 0, \text{ where } d_0, \ldots, d_m \text{ are integers}, \tag{3}$$

necessarily implies that the numbers d_j are equal to zero).

We shall explain the difficulty that arises with a linear dependence between the numbers $\pi, \alpha_1, \ldots, \alpha_m$. Let there be d_j, not all equal to zero, such that Eq. (3) holds. Then, clearly,

$$\kappa_{p'q'} = \kappa_{pq} \quad p' = p + d_0\xi; \quad q_j' = q_j + 2d_j\xi$$

[ξ is any integer, κ_{pq} is determined by Eq. (2), and $\kappa_{p'q'}$ has an analogous sense].

In the language of quantum mechanics the zeroth-approximation eigenvalue (2) is degenerate; different eigenfunctions correspond to the same eigenvalues $\kappa_{pq} = \kappa_{p'q'}$. In the degenerate case, of course, it is more difficult to find the corrections than in the nondegenerate case.

We seek the asymptotic decomposition of the eigenfunction in the form

$$u = e^{i\kappa_{pq} s} \sum_{j=0}^{\infty} \mathcal{U}_{pq}^{(j)} \kappa_{pq}^{-\frac{j}{2}}; \quad \mathcal{U}_{pq}^{(j)} = \mathcal{U}_{pq}^{(j)}(s, \mu_1, \ldots, \mu_m). \tag{4}$$

Here κ_{pq} is the zeroth-approximation eigenvalue (2) (often in the interest of space we shall drop the subscript p), and $e^{i\kappa_q s}\mathcal{U}_q^{(0)}$ is the zeroth-approximation eigenfunction:

$$e^{i\kappa_q s}\mathcal{U}_q^{(0)} = e^{i\kappa_q s} \Lambda_1^{*q_1} \ldots \Lambda_m^{*q_m} \mathcal{U}_0^{(0)},$$

$$\mathcal{U}_0^{(0)} = \frac{1}{\sqrt{\det z}} e^{\frac{i}{2}(z' z^{-1} \vec{\mu}, \vec{\mu})}. \tag{5}$$

We shall assume that the asymptotic decomposition of the eigenvalue κ has a power series form:

$$\kappa = \kappa_q + \sum_{j=1}^{\infty} \frac{\delta_j}{\kappa_q^{j/2}}. \tag{6}$$

Substituting the series (4) and (6) into Eq. (1) and setting the coefficients of like powers of κ_q equal to zero, we arrive at the recursive sequence of equations

$$L_0 \mathcal{U}_q^{(0)} = 0,$$

$$L_0 \mathcal{U}_q^{(1)} = -L_1 \mathcal{U}_q^{(0)},$$

$$\cdots\cdots\cdots\cdots$$

$$L_0 \mathcal{U}_q^{(\ell)} = -L_1 \mathcal{U}_q^{(\ell-1)} - \cdots - L_\ell \mathcal{U}_q^{(0)}.$$

$$\cdots\cdots\cdots\cdots\cdots\cdots \tag{7}$$

Here L_0 is the "parabolic" operator analyzed in [1]:

$$L_0 = 2i \frac{\partial}{\partial s} + \sum_{j=1}^{m} \frac{\partial^2}{\partial \mu_j^2} - (\mathcal{K}(s)\vec{\mu}, \vec{\mu}), \tag{8}$$

and the operators L_j have the form

$$L_j = \overset{0}{R}_j \frac{\partial}{\partial s} + \overset{\infty}{R}_j \frac{\partial^2}{\partial s^2} + \sum_{\tau=1}^{m} \overset{\tau}{R}_{j+1} \frac{\partial}{\partial \mu_\tau} + \sum \overset{n\tau}{R}_j \frac{\partial^2}{\partial \mu_n \partial \mu_\tau} + R_{j+2} + \Psi_j, \tag{9}$$

where $R_j, \ldots$ are polynomials in μ_j with coefficients infinitely differentiable with respect to s. The degrees of these polynomials do not exceed their subscript values. Expression (9) is easily deduced if the coefficients of Eq. (1) are decomposed into formal power series in y_j. We say that the polynomial $R(\vec{\mu})$ is even (odd) if $R(-\vec{\mu}) \equiv R(\vec{\mu})$ $[R(-\vec{\mu}) \equiv -R(\vec{\mu})]$. It can be verified by direct computation that the evenness or oddness of the polynomials R is the same as for their subscripts (i.e., the subscript and the polynomials are either both even or both odd at the same time).

Also, Ψ_j in Eq. (9) is a number defined by the formula

$$\Psi_j = \delta_j + \delta_1 \delta_{j-3} + \cdots + \delta_{j-3}\delta_1 + \delta_j = 2\delta_j + 2\delta_1\delta_{j-3} + \cdots \tag{10}$$

Using the explicit form of the operators Λ_j and Λ_j^* (see [1]):

$$\Lambda_j = \frac{1}{i}(\vec{z}_j, \nabla_\mu) - \left(\frac{d\vec{z}_j}{ds}, \vec{\mu}\right),$$

$$j = 1, 2, \ldots, m,$$

$$\Lambda_j^* = \frac{1}{i}(\vec{z}_j^*, \nabla_\mu) - \left(\frac{d\vec{z}_j^*}{ds}, \vec{\mu}\right),$$

we can represent the operators of multiplication by μ_i and differentiation with respect to μ_i as linear combinations:

$$\mu_i = \sum_{j=1}^{m} A_j^{(i)} \Lambda_j + B_j^{(i)} \Lambda_j^*$$

$$\frac{\partial}{\partial \mu_i} = \sum_{j=1}^{m} (C_j^{(i)} \Lambda_j + D_j^{(i)} \Lambda_j^*).$$

Inserting these expressions into the formula for the operator L_j, we obtain an expression for that operator in the form of a polynomial in the operators $\frac{\partial}{\partial s}$, Λ_i, and Λ_i^* with coefficients depending on s.

Let us assume that:

1) all the numbers δ_j have been found for $j < l$;

2) $U_q^{(j)}$, $j < l$, have the form of polynomials in $\vec{\mu}$ with coefficients infinitely differentiable with respect to s and multiplied by $U_0^{(0)}$;

3) $\exp(i K_q s) U_q^{(j)}$ is L-periodic ($j < l$).

The latter requirement guarantees uniqueness of the eigenfunction in any approximation. For $j = 0$ this requirement is met (see [1]). Substituting $U_q^{(j)}$ ($j < l$) into the ($l+1$)st equation of the system (7) and making use of the properties 1), 2), and 3), the commutation relations for the operators Λ_j, and Λ_j^*, and the identities $\Lambda_j U_0^{(0)} = 0$ (see [1]) to find $U_q^{(l)}$, we obtain the equation

$$\mathcal{L}_0 U_q^{(l)} = \sum_{(\tau)} E_{q\tau}^{(l)}(s) \Lambda_1^{*\tau_1} \cdots \Lambda_m^{*\tau_m} U_0^{(0)}, \tag{11}$$

where $E_{q\tau}^{(l)}(s)$ are infinitely-differentiable functions satisfying the conditions

$$E_{q\tau}^{(l)}(s+L) = E_{q\tau}^{(l)}(s) \cdot \exp(i(K_\tau - K_q)L),$$

$$K_\tau L = 2\pi p + \sum_{j=1}^{m} (\tau_j + \tfrac{1}{2})\alpha_j ; \quad K_q L = 2\pi p + \sum_{j=1}^{m} (q_j + \tfrac{1}{2})\alpha_j . \tag{12}$$

All the functions $E_{q\tau}^{(l)}$ are uniquely determined by the numbers δ_j, $j < l$, and the functions $U_q^{(j)}$ ($j < l$), except for $E_{qq}^{(l)}$. A straightforward computation reveals that

$$E_{qq}^{(l)}(s) = \Psi_l + \mathcal{F}_{qq}^{(l)}(s); \quad \Psi_l = \delta_l + \delta_1 \delta_{l-1} + \cdots + \delta_{l-1}\delta_1 + \delta_l ,$$

where $\mathcal{F}_{qq}^{(l)}(s)$ is a known L-periodic function.

We seek $U_q^{(l)}$ in the form

$$U_q^{(l)} = \sum_{(\tau)} e_{q\tau}^{(l)}(s) \Lambda_1^{*\tau_1} \cdots \Lambda_m^{*\tau_m} U_0^{(0)} . \tag{14}$$

Using Eq. (11) and the fact that the functions

$$U_\tau^{(0)} = \Lambda_1^{*\tau_1} \cdots \Lambda_m^{*\tau_m} U_0^{(0)} \tag{15}$$

satisfy the equation $\mathcal{L}_0 U = 0$, we readily deduce the equations

$$2i \frac{d}{ds} e^{(\ell)}_{q\tau} = E^{(\ell)}_{q\tau}. \tag{16}$$

In order to guarantee the L-periodicity of the functions $\exp(i\kappa_q s) U^{(\ell)}_q$ it is sufficient that the functions $e^{(\ell)}_{q\tau}$ satisfy relations analogous to relations (14):

$$e^{(\ell)}_{q\tau}(s+L) = e^{(\ell)}_{q\tau}(s) \exp(i(\kappa_\tau - \kappa_q)L). \tag{17}$$

The problem of finding the solution of Eq. (16) subject to condition (17) is a self-adjoint boundary-value problem. In order for it to have a solution that is unique it is necessary and sufficient that the corresponding homogeneous problem have only a zero solution. The proof of this statement is easily realized without resort to the general theory, simply by direct integration of both sides of Eq. (16).

By virtue of the assumed linear independence of $\pi, \alpha_1, \ldots, \alpha_m$ over the ring of integers for $\tau \neq q$ all the boundary-value problems (16)-(17) are uniquely solvable.

For $\tau = q$ the problem (16)-(17) is already described at the eigenvalue. It is seen at once that in order for it to be solvable it is necessary and sufficient that

$$\int_0^L E^{(\ell)}_{qq}\, ds = 0.$$

Substituting here in place of $E^{(\ell)}_{qq}$ its expression (13), we obtain

$$2\delta_\ell \cdot L + (2\delta_1 \delta_{\ell-3} + 2\delta_2 \delta_{\ell-4} + \ldots)L + \int_0^L \mathcal{F}^{(\ell)}_{qq}(s)\, ds = 0, \tag{18}$$

from which δ_ℓ is uniquely determined. All approximations are found successively in this manner.

It is a simple matter to prove by induction that:

1) the corrections $U^{(j)}_q$ can always be chosen so that they are even or odd according as the number $j + |q| = j + \sum_{j=1}^{m} q_j$ is even or odd (i.e., the function $U^{(j)}_q$ and the number $j + |q|$ are either both even or both odd);

2) all the numbers δ_j with odd indices are equal to zero.

Thus, we let the induction hypothesis hold for $j < \ell$; then, as implied by the constructions just described [see expression (9) for the operators L_j and the discussion following Eq. (9)], on the right-hand side of Eq. (11) all the $E^{(\ell)}_{q\tau} (q \neq \tau)$ for which $|\tau| = \tau_1 + \ldots + \tau_m$ and $|q| = \ell + q_1 + \ldots + q_m$ are not both odd at the same time are equal to zero. From this we at once infer assertion 1) for the case of even ℓ.

Now let ℓ be odd. Then the function $\mathcal{F}^{(\ell)}_{qq}$ is equal to zero [see Eq. (13)].

We infer at once from Eq. (18) and the induction hypothesis for odd ℓ that $\delta_\ell = 0$, and, setting $e^{(\ell)}_{qq} = 0$ (which is admissible, since for $\delta_\ell = 0$ the function $E^{(\ell)}_{qq}$ is also equal to zero), we find that the function $U^{(\ell)}_q$ is even or odd according as $\ell + |q|$ is even or odd.

This justifies assertions 1) and 2).

LITERATURE CITED

1. Babich, V. M. Eigenfunctions concentrated in a neighborhood of a closed geodesic, Seminars in Mathematics, Vol. 9: Mathematical Problems in Wave Propagation Theory, Consultants Bureau, New York (1970), pp. 7-26.

CORRECTION TERM FOR THE EIGENFREQUENCIES OF A THREE-DIMENSIONAL RESONATOR WITH NONOPENING MIRRORS

N. V. Svanidze

Consider a double-mirror resonator for which the principal directions of curvature of the mirrors do not open out. The resonator is filled with a homogeneous medium invested with a wave propagation velocity c $(c \equiv 1)$. The equations for the surfaces of mirrors A and B have the following respective forms in a Cartesian coordinate system (X, Y, Z) :

$$A : z = \frac{X^2}{2R_1} + \frac{Y^2}{2R_2} + A_4(X, Y),$$

$$B : z = d - \frac{X^2}{2R_3} - \frac{Y^2}{2R_4} - B_4(X, Y). \tag{1}$$

Here d is the distance between the mirrors, R_1, R_2 and R_3, R_4 are the principal radii of curvature of mirrors A and B, and

$$A_4(XY) = A^{1111}X^4 + A^{1112}X^3Y + A^{1122}X^2Y^2 + A^{1222}XY^3 + A^{2222}Y^4,$$

where $A^{ijlm} = \mathrm{const}$ and the form of $B_4(X, Y)$ is obtained from the form of $A_4(X, Y)$ by substitution of the coefficients A^{ijlm} in place of B^{ijlm}.

The problem of the eigenmodes of the resonator is reduced to the following for the Helmholtz equation:

$$(\Delta + k^2)U = 0, \qquad U\big|_{A,B} = 0. \tag{2}$$

It is well known [1] that the asymptotic representation of the eigenvalues of the resonator has the form

$$d\,k_{qn_1n_2} = q\pi + \left(n_1 + \frac{1}{2}\right)\varphi_1 + \left(n_2 + \frac{1}{2}\right)\varphi_2 + \frac{\alpha}{k} + O\left(\frac{1}{k^2}\right),$$

where q is an integer, $q \gg 1$, and $n_{1,2} = 0, 1 \ldots$.

In the present article we calculate the number α. The calculations are based on M. M. Popov's paper, "Eigenmodes of Multimirror Resonators" [2]. It would be helpful to the reader to be familiar with this paper.

I. Eigenvectors and Eigenvalues of the Matrix

If a certain vector $\vec{A} = (a_1 a_2 a_3 a_4)$ describes a ray incident on mirror B, and $\vec{A}^{(1)}$ describes the ray obtained from the original one by reflection from B and then from mirror A, then in the first approximation for rays near the resonator axis we obtain $\vec{A}^{(1)} = \Gamma \vec{A}$, where Γ is a fourth-order matrix determined by the parameters of the resonator and endowed with the partitioned-diagonal form $\Gamma = [\Gamma_1, \Gamma_2]$, where Γ_1 and Γ_2 are real second-order matrices. It can be shown [2] that $\det \Gamma = 1$ and $\det \Gamma_{1,2} = 1$.

Let the vectors $\vec{A}_1 = (b_2, b_1, 0, 0)$, $\vec{A}_2 = \bar{\vec{A}}_1$, $\vec{A}_3 = (0, 0, c_2, c_1)$, and $\vec{A}_4 = \bar{\vec{A}}_3$ be the eigenvectors, and let the numbers $\lambda_1 = e^{i2\varphi_1}$; $\lambda_2 = \bar{\lambda}_1$, $\lambda_3 = e^{i2\varphi_2}$, and $\lambda_4 = \bar{\lambda}_3$ be the corresponding eigenvalues of the matrix Γ. The vectors $\vec{A}_1$ and $\vec{A}_3$ are normalized by the conditions

$$\begin{cases} \bar{b}_2 b_1 - b_2 \bar{b}_1 = +i, \\ \bar{c}_2 c_1 - c_2 \bar{c}_1 = +i. \end{cases} \tag{3}$$

Expressions for $\Gamma_{1,2}$, $\varphi_{1,2}$, and the moduli squared of the components of the eigenvectors in terms of the resonator parameters are given in the Appendix (A1 = A3).

II. Solution of the Helmholtz Equation (2)

In accordance with the parabolic equation method [3] we transform to the variables $x = \sqrt{k}\, X$, $y = \sqrt{k}\, Y$, and $z = Z$ and represent the function U in the form $U = e^{ikz} V(x, y, z)$. The Helmholtz equation (2) goes over to

$$\left[2i \frac{\partial}{\partial z} + \Delta_{x,y} + \frac{1}{k} \frac{\partial^2}{\partial z^2}\right] V = 0, \tag{4}$$

where

$$\Delta_{xy} = \frac{\partial^2}{\partial x^2} + \frac{\partial^2}{\partial y^2}.$$

We introduce the operators $L_0 \equiv 2i \frac{\partial}{\partial z} + \Delta_{xy}$ and $L_1 \equiv \frac{\partial^2}{\partial z^2}$, whereupon Eq. (4) is rewritten

$$L_0 V + \frac{1}{k} L_1 V = 0. \tag{5}$$

Next we introduce operators Λ_i corresponding to the eigenvalues λ_i $(i = 1, 2, 3, 4)$ of Γ:

$$\begin{cases} \Lambda_1 = (b_1 z + b_2) \frac{\partial}{\partial x} - i b_1 x, \\ \Lambda_2 = (\bar{b}_1 z + \bar{b}_2) \frac{\partial}{\partial x} - i \bar{b}_1 x; \end{cases} \tag{6}$$

$$\begin{cases} \Lambda_3 = (c_1 z + c_2) \frac{\partial}{\partial y} - i c_1 y, \\ \Lambda_4 = (\bar{c}_1 z + \bar{c}_2) \frac{\partial}{\partial y} - i \bar{c}_1 y, \end{cases} \tag{7}$$

as well as the function $v = \exp\left[\frac{i}{2}(T\vec{x}, \vec{x}) - \frac{1}{2}\int_0^z \operatorname{sp} T(z')\, dz'\right]$, which satisfies the equation

$$L_0 \vartheta \equiv 0. \tag{8}$$

Here $T(z)$ is a symmetric matrix whose elements depend on z. Following are the properties of the operators Λ_i:

$$\begin{aligned}
&1.\quad \Lambda_{1,3}\vartheta \equiv 0, \qquad \Lambda_{2,4}\vartheta \not\equiv 0.
\end{aligned}$$

Consequently, Λ_1 and Λ_3 may be called "destruction" operators, and Λ_2 and Λ_4 may be called "creation" operators.

$$\begin{aligned}
&2.\quad L_0\Lambda_i^n = \Lambda_i^n L_0 \qquad i=1,2,3,4 \qquad n=0,1,\ldots\\
&3.\quad \Lambda_p\Lambda_{p+1} - \Lambda_{p+1}\Lambda_p = -1 \qquad p=1,3,
\end{aligned} \tag{9}$$

all other operators commuting with one another:

$$\Lambda_{1,2}\Lambda_{3,4} = \Lambda_{3,4}\Lambda_{1,2}.$$

We regard expressions (6) and (7) as systems of two equations in the unknowns $x, \frac{\partial}{\partial x}$ and $y, \frac{\partial}{\partial y}$. Solving them formally and taking account of relations (3), we obtain

$$\left.\begin{aligned}
x &= (\bar{b}_1 z + \bar{b}_2)\Lambda_1 - (b_1 z + b_2)\Lambda_2,\\
\frac{\partial}{\partial x} &= i\bar{b}_1\Lambda_1 - i b_1\Lambda_2,
\end{aligned}\right\} \tag{10}$$

$$\left.\begin{aligned}
y &= (\bar{c}_1 z + \bar{c}_2)\Lambda_3 - (c_1 z + c_2)\Lambda_4,\\
\frac{\partial}{\partial y} &= i\bar{c}_1\Lambda_3 - i c_1\Lambda_4.
\end{aligned}\right\} \tag{11}$$

We seek the solution of Eq. (4) in the form

$$V = \Lambda_2^{n_1}\Lambda_4^{n_2}\vartheta + \frac{1}{k}\sum_{\sigma_{n_1 n_2}} C_{m_1 m_2 n_1 n_2}(z)\Lambda_2^{n_1+m_1}\Lambda_4^{n_2+m_2}\vartheta + O\left(\frac{1}{k^2}\right), \tag{12}$$

where $C_{m_1 m_2 n_1 n_2}(z)$ are as yet unknown functions of z, and $\sigma_{n_1 n_2}$ is the following set of pairs of integers m_1, m_2

$$\sigma_{n_1 n_2} = \left\{ m_1, m_2 : \begin{array}{c} -4 \le m_1, m_2 \le 4 \\ n_1 + m_1 \ge 0 \\ n_2 + m_2 \ge 0 \end{array} \right\} \qquad n_1, n_2 = 0, 1, \ldots. \tag{13}$$

Let $U_{n_1 n_2} \equiv \Lambda_2^{n_1}\Lambda_4^{n_2}\vartheta$; using the properties of the operators Λ_i (9) and Eq. (8), we readily show that

$$L_0 U_{n_1 n_2} \equiv 0. \tag{14}$$

Substituting expression (12) into Eq. (5) and bearing (14) in mind, we have

$$\left(L_0 + \frac{1}{k}L_1\right)V = \frac{1}{k}\left[\sum_{\sigma_{n_1 n_2}} L_0 C_{m_1 m_2 n_1 n_2}(z) U_{n_1+m_1, n_2+m_2} + L_1 U_{n_1 n_2}\right] + O\left(\frac{1}{k^2}\right) = 0. \tag{15}$$

The coefficients $C_{m_1 m_2 n_1 n_2}(z)$ are determined from the condition that the term with $\frac{1}{k}$ is equal to zero in (15).

We now consider each term contained in the brackets of Eq. (15).

1. $L_0 C_{m_1 m_2 n_1 n_2}(z) U_{n_1+m_1, n_2+m_2} = 2i\left(\frac{\partial}{\partial z} C_{m_1 m_2 n_1 n_2}(z)\right) U_{n_1+m_1, n_2+m_2}.$

Here we have used the explicit form of the operator L_0 and relation (14).

2. If Eq. (14) is rewritten in the form

$$\frac{\partial}{\partial z} U_{n_1 n_2} = \frac{i}{2} \Delta_{xy} U_{n_1 n_2}, \tag{16}$$

and both sides are then differentiated with respect to z and (16) is taken into account, we obtain

$$L_1 U_{n_1 n_2} = -\frac{1}{4} \Delta^2_{xy} U_{n_1 n_2}$$

We introduce the operator $\tilde{L}_1 \equiv -\frac{1}{4} \Delta^2_{xy}$, which is reduced to the following by the replacement of $\frac{\partial}{\partial x}$ and $\frac{\partial}{\partial y}$ according to Eqs. (10) and (11):

$$\tilde{L}_1 = \sum_{\substack{i+j+r+s=0,2,4 \\ i,j,r,s \geq 0}} H_{ijrs} \Lambda_2^i \Lambda_4^j \Lambda_1^r \Lambda_3^s, \tag{17}$$

where H_{ijrs} are constants expressed in terms of the components of the eigenvectors of Γ (see A5 in the Appendix). Consequently, the action of the operator L_1 on the function $U_{n_1 n_2}$ is equivalent to the action of $\tilde{L}_1$.

With due regard for parts 1 and 2 above and A4 of the Appendix, we arrive at the following equation for the unknown coefficients $C_{m_1 m_2 n_1 n_2}(z)$:

$$\sum_{G_{n_1 n_2}} \left[\frac{d}{dz} C_{m_1 m_2 n_1 n_2}(z) - D_{m_1 m_2 n_1 n_2} \right] U_{n_1+m_1, n_2+m_2} = 0, \tag{18}$$

where $D_{m_1 m_2 n_1 n_2}$ are constants expressed in terms of the coefficients H_{ijrs} (see A6).

From Eq. (18) we readily find the explicit form of the coefficients $C_{m_1 m_2 n_1 n_2}(z)$:

$$C_{m_1 m_2 n_1 n_2}(z) = D_{m_1 m_2 n_1 n_2} z + C_{m_1 m_2 n_1 n_2}(0). \tag{19}$$

The constants $C_{m_1 m_2 n_1 n_2}(0)$ are determined from the boundary conditions.

III. Boundary Conditions

Consider a wave $\tilde{u} = e^{ikz} U_{n_1 n_2}$ traveling from mirror A to B, on the surface of mirror A:

$$\tilde{u}\big|_A = (e^{ikz})_A (U_{n_1 n_2})_A = e^{iA_2(x,y)} \left[(U_{n_1 n_2})_{z=0} + \frac{1}{k}\left(iA_4(x,y) U_{n_1 n_2} + A_2(x,y) \frac{\partial}{\partial z} U_{n_1 n_2} \right)_{z=0} \right] + O\left(\frac{1}{k^2}\right), \tag{20}$$

where $A_2(x,y) \equiv \frac{x^2}{2R_1} + \frac{y^2}{2R_2}$. We introduce two operators L_2 and L_3 as follows.

1. The operator L_2 is obtained by the formal change of the variables x and y in $iA_4(x,y)$ according to Eqs. (10) and (11):

$$L_2 = \sum_{\substack{i+j+r+s=0,2,4 \\ i,j,r,s \geq 0}} \Pi_{ijrs}(z) \Lambda_2^i \Lambda_4^j \Lambda_1^r \Lambda_3^s, \tag{21}$$

where the coefficients $\Pi_{ijrs}(z)$ are expressed in terms of the components of the eigenvectors of Γ and are functions of z (see A5). Multiplication of $U_{n_1 n_2}$ by $iA_4(x,y)$ is equivalent to the action of the operator L_2 on $U_{n_1 n_2}$.

2. Bearing Eq. (16) in mind, we see that

$$\left(A_2(x,y)\frac{\partial}{\partial z}\right)U_{n_1n_2}=\left(A_2(x,y)\frac{i}{2}\Delta_{xy}\right)U_{n_1n_2},$$

and by analogy with (21) we change x, y, $\frac{\partial}{\partial x}$, and $\frac{\partial}{\partial y}$ according to Eqs. (10) and (11) to obtain the operator L_3:

$$L_3=\sum_{\substack{i+j+\tau+s=0,2,4\\ i,j,\tau,s\geq 0}} M_{ij\tau s}(z)\Lambda_2^i\Lambda_4^j\Lambda_1^\tau\Lambda_3^s, \tag{22}$$

where $M_{ij\tau s}(z)$ are expressed in terms of the components of the eigenvectors of Γ and are functions of z (see A5), and

$$\left[A_2(x,y)\frac{\partial}{\partial z}\right]U_{n_1n_2}=L_2U_{n_1n_2}.$$

With regard for parts 1 and 2 above and A4 in the Appendix, we rewrite the value of $\tilde{u}$ at the boundary A as follows:

$$\tilde{u}\Big|_A=e^{iA_2(x,y)}\left[U_{n_1n_2}+\frac{1}{\kappa}\left(\sum_{\sigma_{n_1n_2}}N_{m_1m_2n_1n_2}(z)\Lambda_2^{n_1+m_1}\Lambda_4^{n_2+m_2}\vartheta\right)\right]_{z=0}+O\left(\frac{1}{\kappa^2}\right), \tag{23}$$

where $N_{m_1m_2n_1n_2}(z)$ are functions of z expressed in terms of $M_{ij\tau s}$ and $\Pi_{ij\tau s}$ (see A7).

The same is true for the function $\tilde{u}$ at the boundary B:

$$\tilde{u}\Big|_B=e^{i\kappa d-iB_2(x,y)}\left[U_{n_1n_2}+\frac{1}{\kappa}\left(\sum_{\sigma_{n_1n_2}}\tilde{N}_{m_1m_2n_1n_2}(z)U_{n_1+m_1,n_2+m_2}\right)\right]_{z=d}+O\left(\frac{1}{\kappa^2}\right), \tag{24}$$

where $B_2(x,y)\equiv\frac{x^2}{2R_3}+\frac{y^2}{2R_4}$ and $\tilde{N}_{m_1m_2n_1n_2}(z)$ are functions of z (see A8).

Next we consider the boundary conditions up to terms of order $O\left(\frac{1}{\kappa^2}\right)$. Everything that applies to the wave traveling in the direction of mirror B is assigned to the index $(+)$:

$$u^{(+)}_{n_1n_2}=\exp\left\{i\kappa z-\frac{1}{2}\int_0^z \mathrm{sp}\,T^{(+)}(z')dz'\right\}U^{(+)}_{n_1n_2},$$

$$U^{(+)}_{n_1n_2}=[\Lambda_2^{(+)}]^{n_1}[\Lambda_4^{(+)}]^{n_2}U_0^{(+)},\qquad U_0^{(+)}=e^{\frac{i}{2}(T^{(+)}\vec{x},\vec{x})}.$$

The wave propagating in the direction of mirror A is obtained from the wave $u^{(+)}_{n_1n_2}$ by the substitution of $(-i)$ for i. We denote the result by $u^{(-)}_{n_1n_2}$:

$$u^{(-)}=\exp\left\{-i\kappa z-\frac{1}{2}\int_0^z \mathrm{sp}\,T^{(-)}(z')dz'\right\}U^{(-)}_{n_1n_2},$$

$$U^{(-)}_{n_1n_2}=[\Lambda_2^{(-)}]^{n_1}[\Lambda_4^{(-)}]^{n_2}U_0^{(-)},\quad U_0^{(-)}=e^{-\frac{i}{2}(T^{(-)}\vec{x},\vec{x})}.$$

Now the boundary conditions (2) are written

$$\left.\begin{aligned} (\mathcal{U}^{(+)}-\mathcal{U}^{(-)})\big|_A &= O\left(\tfrac{1}{\kappa^2}\right), \\ (\mathcal{U}^{(+)}-\mathcal{U}^{(-)})\big|_B &= O\left(\tfrac{1}{\kappa^2}\right), \end{aligned}\right\} \tag{25}$$

where

$$\left.\begin{aligned} \mathcal{U}^{(\pm)}\big|_A &= \Big[u^{(\pm)}_{n_1 n_2} + \frac{1}{\kappa}\sum_{\sigma_{n_1,n_2}}\Big(C^{(\pm)}_{m_1 m_2 n_1 n_2}(z) + N^{(\pm)}_{m_1 m_2 n_1 n_2}(z)\Big)u^{(\pm)}_{n_1+m_1,\,n_2+m_2}\Big]_{z=0} + O\left(\tfrac{1}{\kappa^2}\right), \\ \mathcal{U}^{(\pm)}\big|_B &= \Big[u^{(\pm)}_{n_1 n_2} + \frac{1}{\kappa}\sum_{\sigma_{n_1,n_2}}\Big(C^{(\pm)}_{m_1 m_2 n_1 n_2}(z) + \tilde{N}^{(\pm)}_{m_1 m_2 n_1 n_2}(z)\Big)u^{(\pm)}_{n_1+m_1,\,n_2+m_2}\Big]_{z=d} + O\left(\tfrac{1}{\kappa^2}\right). \end{aligned}\right. \tag{26}$$

We seek the wave number κ in the form

$$\kappa = \kappa^*_{q n_1 n_2} + \frac{\alpha}{\kappa} + O\left(\tfrac{1}{\kappa^2}\right), \tag{27}$$

where $\kappa^*_{q n_1 n_2}$ and α are as yet unknown numbers.

We substitute expressions (26) and (27) into the boundary conditions (25). Setting terms of order $O(1)$ equal to zero, we obtain

$$\left.\begin{aligned} \big(u^{(+)}_{n_1 n_2} - u^{(-)}_{n_1 n_2}\big)_{z=0} &= 0, \\ \big(u^{(+)}_{n_1 n_2} - u^{(-)}_{n_1 n_2}\big)_{z=d} &= 0, \end{aligned}\right\} \tag{28}$$

where

$$\begin{aligned} u^{(\pm)}_{n_1 n_2}\big|_{z=0} &= \exp[\pm i A_2(x,y)]\,(U^{(\pm)}_{n_1 n_2})_{z=0}, \\ u^{(\pm)}_{n_1 n_2}\big| &= \exp\Big[\pm i\kappa d \mp B_2(x,y) - \tfrac{1}{2}\int_0^d \mathrm{sp}\,T^{(\pm)}(z')\,dz'\Big](U^{(\pm)}_{n_1 n_2})_{z=d} \end{aligned} \tag{29}$$

From conditions (28) we find an expression for $\kappa^*_{q n_1 n_2}$:

$$d\kappa^*_{q n_1 n_2} = q\pi + (n_1 + \tfrac{1}{2})\varphi_1 + (n_2 + \tfrac{1}{2})\varphi_2,$$

where q is an integer, $q \gg 1$, and $n_1, n_2 = 0, 1, \ldots$.

We now set terms of order $O(\frac{1}{\kappa})$ equal to zero. We obtain a system of linear equations in the unknowns $C^{(+)}_{m_1 m_2 n_1 n_2}(0)$ and $C^{(-)}_{m_1 m_2 n_1 n_2}(0)$ $(m_1, m_2 \in \sigma_{n_1 n_2})$, which decomposes into a series of linear systems of two equations for $C^{(+)}_{m_1 m_2 n_1 n_2}(0)$ and $C^{(-)}_{m_1 m_2 n_1 n_2}(0)$, now with a fixed pair of indices $m_1, m_2 \in \sigma_{n_1 n_2}$.

If $m_1 \neq 0$ and $m_2 \neq 0$, then for solvability of the linear systems of equations in $C^{(+)}_{m_1 m_2 n_1 n_2}(0)$ and $C^{(-)}_{m_1 m_2 n_1 n_2}(0)$ it is necessary that the following conditions be met:

$$m_1\varphi_1 + m_2\varphi_2 \neq n\pi$$
$$m_1 m_2 \in \mathfrak{S}_{n_1 n_2}, \quad m_1 \neq 0, \quad m_2 \neq 0, \tag{30}$$

where n is an integer. In other words, in addition to the conditions of resonator stability in the first approximation, it is required that restrictions be imposed on the resonator parameters in connection with conditions (30).

For $m_1 = m_2 = 0$, we arrive at a linear system of two equations for the determination of the unknowns $C^{(+)}_{00n_1n_2}(0)$ and $C^{(-)}_{00n_1n_2}(0)$.

The determinant of this system is always equal to zero, and the number α is determined from the solvability requirement on the system:

$$\alpha = \frac{1}{2id}\left[\left(\mathcal{N}^{(+)}_{00n_1n_2} - \mathcal{N}^{(-)}_{00n_1n_2}\right) - \left(\tilde{\mathcal{N}}^{(+)}_{00n_1n_2} - \tilde{\mathcal{N}}^{(-)}_{00n_1n_2}\right) - d\left(\mathcal{D}^{(+)}_{00n_1n_2} - \mathcal{D}^{(-)}_{00n_1n_2}\right)\right]. \tag{31}$$

If the coefficients $\mathcal{N}^{(\pm)}_{00n_1n_2}(0)$, $\tilde{\mathcal{N}}^{(\pm)}_{00n_1n_2}(d)$, $\mathcal{D}^{(\pm)}_{00n_1n_2}$ are calculated explicitly using the formulas of the Appendix, the expression for α in terms of the resonator parameters acquires the form

$$\alpha_{n_1n_2} = \frac{1}{64d}(t_1^2+9)\left(\frac{1}{R_1}+\frac{1}{R_3}\right) + \frac{1}{64d}(t_2^2+9)\left(\frac{1}{R_2}+\frac{1}{R_4}\right) +$$

$$+\frac{3}{64}(t_1^2+1)\frac{1}{R_1+R_3-d}\left[\left(8A^{1111}-\frac{1}{R_1^3}\right)R_1^2\frac{d-R_3}{d-R_1} + \left(8B^{1111}-\frac{1}{R_3^3}\right)\frac{d-R_1}{d-R_3}\right] +$$

$$+\frac{1}{16}t_1t_2\frac{1}{\sqrt{(R_1+R_3-d)(R_2+R_4-d)}}\left[\left(4A^{1122}-\frac{1}{dR_1R_2}\left\langle 1+\frac{1}{2}\frac{d-R_1}{d-R_3}+\frac{1}{2}\frac{d-R_2}{d-R_4}\right\rangle\right)R_1R_2\sqrt{\frac{(d-R_3)(d-R_4)}{(d-R_1)(d-R_2)}} +\right.$$

$$\left.+\left(4B^{1122}-\frac{1}{dR_3R_4}\left\langle 1+\frac{1}{2}\frac{d-R_3}{d-R_1}+\frac{1}{2}\frac{d-R_4}{d-R_2}\right\rangle\right)R_3R_4\sqrt{\frac{(d-R_1)(d-R_2)}{(d-R_3)(d-R_4)}}\right] +$$

$$+\frac{3}{64}(t_2^2+1)\frac{1}{R_2+R_4-d}\left[\left(8A^{2222}-\frac{1}{R_2^3}\right)R_2^2\frac{d-R_4}{d-R_2} + \left(8B^{2222}-\frac{1}{R_4^3}\right)R_4^2\frac{d-R_2}{d-R_4}\right], \tag{32}$$

$$t_1 = 2n_1+1, \quad t_2 = 2n_2+1, \quad n_1, n_2 = 0, 1, 2.$$

As apparent from (32), the formula for α does not involve terms containing the coefficients A^{1112}, A^{1222} or B^{1112}, B^{1222} from the equations for the surfaces of mirrors A and B. It is important to note that if one of the conditions (30) were not fulfilled (spectral degeneracy case; see [1]), the formula for α would not longer be valid.

For example, if we consider an oblate ellipsoid of revolution as the resonator, then, as shown by a formula derived by S. Yu. Slavyanov through the separation of variables, expression (32) yields an incorrect value of α.

Equation (32) for α exactly coincides with the formula derived by V. F. Lazutkin for the two-dimensional case [4].

APPENDIX

A1.

$$\Gamma_1 = \left\| \begin{matrix} 1-\frac{2d}{R_3} & 2d-\frac{2d^2}{R_3} \\ \frac{4d}{R_1R_3}-\frac{2}{R_1}-\frac{2}{R_3} & \frac{4d^2}{R_1R_3}-\frac{4d}{R_1}-\frac{2d}{R_3}+1 \end{matrix} \right\|$$

A2.

$$\varphi_1 = \arccos \theta \sqrt{\left(1-\frac{d}{R_1}\right)\left(1-\frac{d}{R_3}\right)},$$

where $\theta = \operatorname{sign}(R_1+R_3-2d)$.

A3.

$$|b_2|^2 = \frac{R_1}{2}\left[\frac{d}{R_1+R_3-d}\,\frac{d-R_3}{d-R_1}\right]^{1/2},$$

$$|b_1 d + b_2|^2 = \frac{R_3}{2}\left[\frac{d}{R_1+R_3-d}\,\frac{d-R_1}{d-R_3}\right]^{\frac{1}{2}},$$

$$d|b_1|^2 = \frac{|b_1 d + b_2|}{R_3} + \frac{|b_2|^2}{R_1}.$$

The explicit form of Γ_2, φ_2, $|c_2|^2$, $|c_1|^2$, and $|c_1 d + c_2|^2$ is obtained from the above expressions with the replacement of b_1, b_2 by c_1, c_2 and of the subscripts $i = 1,3$ on R_i by $i = 2,4$.

A4. Commutation rule for the operators Λ_p^m and Λ_{p+1}^n, $p = 1,3$, $m,n = 0,1,\ldots$:

$$\Lambda_p^m \Lambda_{p+1}^n = \sum_{j=0}^{m} (-1)^j C_m^j \frac{n!}{(n-j)!} \Lambda_{p+1}^{n-j} \Lambda_p^{m-j} \quad \text{for } m \leq n; \text{ for}$$

$$\Lambda_p^m \Lambda_{p+1}^n = \sum_{j=0}^{n} (-1)^j C_n^j \frac{m!}{(m-j)!} \Lambda_{p+1}^{n-j} \Lambda_p^{m-j} \quad \text{for } m \geq n. \text{ for}$$

A5. Expressions (17), (21), and (22) for the operators $\tilde{L}_1$, L_2, and L_3 are easily written in explicit form with the aid of the following formulas.

Let

$$l = l_1\Lambda_1 - l_2\Lambda_2, \quad h = h_1\Lambda_1 - h_2\Lambda_2,$$

$$f = f_3\Lambda_3 - f_4\Lambda_4,$$

where l_1, l_2, h_1, h_2, f_3, and f_4 are constants for functions of z. It is apparent from the properties (9) of the operators Λ_i that $fl = lf$, $fh = hf$;

$$h^4 = \sum_{i+\tau=4} (-1)^i \frac{4!}{i!\tau!} h_1^\tau h_2^i \Lambda_2^i \Lambda_1^\tau + 6h_1h_2 \sum_{i+\tau=2} (-1)^i \frac{2!}{i!\tau!} h_1^\tau h_2^i \Lambda_2^i \Lambda_1^\tau + 3h_1^2 h_2^2.$$

An analogous formula is obtained for f^4;

$$h^2 l^2 = \sum_{\substack{i+\tau=2 \\ j+s=2}} (-1)^{i+j} \frac{2!}{i!\tau!} \frac{2!}{j!s!} h_1^\tau h_2^i l_1^s l_2^j \Lambda_2^{i+j} \Lambda_1^{\tau+s} + h_1h_2 \sum_{j+s=2} (-1)^j \frac{2!}{j!s!} l_1^s l_2^j \Lambda_2^j \Lambda_1^s + l_1 l_2 \sum_{i+\tau=2} (-1)^i \frac{2!}{i!\tau!} h_1^\tau h_2^i \Lambda_2^i \Lambda_1^\tau +$$

$$+4h_1 l_2 \sum_{\substack{i+r=1\\ j+s=1}} (-1)^{i+j} h_1^r h_2^i l_1^s l_2^j \Lambda_2^{i+j} \Lambda_1^{r+s} + 2h_1^2 l_2^2 + h_1 h_2 l_1 l_2 ;$$

$$h^2 f^2 = \sum_{\substack{i+r=2\\ j+s=2}} (-1)^{i+j} \frac{2!}{i!r!} \frac{2!}{j!s!} h_1^r h_2^i f_3^s f_4^j \Lambda_2^i \Lambda_4^j \Lambda_1^r \Lambda_3^s +$$

$$+ h_1 h_2 \sum_{j+s=2} (-1)^j \frac{2!}{j!s!} f_3^s f_4^j \Lambda_4^j \Lambda_3^s + f_3 f_4 \sum_{i+r=2} (-1)^i \frac{2!}{i!r!} h_1^r h_2^i \Lambda_2^i \Lambda_1^r + h_1 h_2 f_3 f_4 ;$$

$$h^3 f = \sum_{\substack{i+r=3\\ j+s=1}} \frac{3!}{i!r!} (-1)^{i+j} h_1^r h_2^i f_3^s f_4^j \Lambda_2^i \Lambda_4^j \Lambda_1^r \Lambda_3^s + 3h_1 h_2 \sum_{\substack{i+r=3\\ j+s=1}} (-1)^{i+j} h_1^r h_2^i f_3^s f_4^j \Lambda_2^i \Lambda_4^j \Lambda_1^r \Lambda_3^s .$$

A6. $$D_{m_1 m_2 n_1 n_2} = \frac{i}{2} \sum_{\Omega_{m_1 m_2 n_1 n_2}} (-1)^{r+s} \frac{n_1!}{(n_1-r)!} \frac{n_2!}{(n_2-s)!} H_{ijrs} ,$$

where $\Omega_{m_1 m_2 n_1 n_2}$ is the following set of integers:

$$\Omega_{m_1 m_2 n_1 n_2} = \left\{ i,j,r,s : \begin{matrix} i+j+r+s=0,2,4 \\ i,j,r,s \geq 0 \end{matrix} ; \begin{matrix} i-r=m_1 \\ j-s=m_2 \end{matrix} ; \begin{matrix} r \leq n_1 \\ s \leq n_2 \end{matrix} \right\} .$$

A7. $$N_{m_1 m_2 n_1 n_2}(z) = \sum_{\Omega_{m_1 m_2 n_1 n_2}} (-1)^{r+s} \frac{n_1!}{(n_1-r)!} \frac{n_2!}{(n_2-s)!} \left[M_{ijrs}(z) + \Pi_{ijrs}(z) \right] .$$

A8. The coefficients $\tilde{N}_{m_1 m_2 n_1 n_2}(z)$ can be obtained from $N_{m_1 m_2 n_1 n_2}(z)$ by the formal replacement of R_1, R_2, and A^{ijlm} by $(-R_3)$, $(-R_4)$, and $(-B^{ijlm})$.

LITERATURE CITED

1. Lazutkin, V. F., Spectral degeneracy and "small denominators" in the asymptotic representation of eigenfunctions of the "bouncing ball" type, Vest. Leningrad. Univ., No. 7 (1969).
2, Popov, M. M., Eigenmodes of multimirror resonators, Vest. Leningrad. Univ. (in press).
3. Lazutkin, V. F., The parabolic equation and asymptotic represention of the eigenfunctions of the Helmholtz equation for three-dimensional domains, Vest. Leningrad. Univ., No. 22, pp. 52-57 (1965).
4. Lazutkin, V. F., Formula for the eigenfrequencies of a nonconfocal resonator with cylindrical mirrors, taking into account aberration of the mirrors, Opt. i Spektrosk., 24:453-454 (1968).

ASYMPTOTIC BEHAVIOR OF THE EIGENFUNCTIONS AND EIGENFREQUENCIES OF A MULTIMIRROR RESONATOR

B. N. Semenov

In the present article we investigate the problem of determining the higher approximations for the eigenfunctions and eigenfrequencies of a multimirror annular resonator filled with an inhomogeneous medium characterized by a wave propagation velocity $c(\vec{\tau})$. Their zeroth approximation has been obtained in [2].

We define rays in the inhomogeneous space as the extremals of the geometrical-optic functional

$$\mathcal{J} = \int_{M_0}^{M_1} \frac{d\sigma}{c(\vec{\tau})}.$$

Let there be a system of mirrors (see Fig. 1) situated so that there exists a ray emanating from some point A and, after reflection from N mirrors, transforming into itself ($d_1, d_2, \ldots, d_N$ are the points of reflection of the ray). We call the indicated ray the axis of the resonator and refer to the segment $d_{i-1}\, d_i$ of the axis as the axis of the i-th arm of the resonator.

At each mirror S_i we introduce a Cartesian coordinate system ξ_{1i}, ξ_{2i}, ξ_{3i} as follows: At the point of incidence of the ray we direct the ξ_{1i} axis along the bisector of the angle formed by the adjacent arms of the resonator, we make the ξ_{2i} axis perpendicular to the plane of reflection of the ray, and we direct the ξ_{3i} axis along the normal to the surface of the mirror into the resonator. The equation for the mirror surface S_i in this coordinate system is written in the form

$$\xi_{3i} = \sum_{m+n \geq 2} g_{mn}^{(i)} \xi_{1i}^{m} \xi_{2i}^{n}. \tag{1}$$

Our aim is to formulate the high-frequency asymptotic behavior of the eigenmodes of the resonator (ω is a large fixed parameter). The problem reduces to the formulation of solutions to the equation

$$\Delta U + \frac{\omega^2}{c^2(\vec{\tau})} U = 0, \tag{2}$$

such as will satisfy the following boundary conditions on the mirrors:

$$U\big|_{S_j} = 0 \qquad (\kappa = 1, 2, \ldots, N) \tag{3}$$

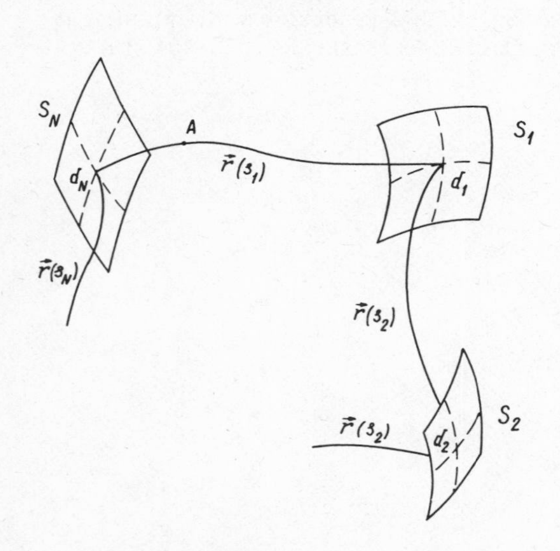

Fig. 1.

and the following condition of concentration in the vicinity of the resonator axis:

$$\max |U(\vec{r})| \to 0$$

as $R \to \infty$ (where R is the distance from the resonator axis).

We shall assume that the resonator axis is given by the equation $\vec{r} = \vec{r}(s)$, where s is the arc length measured along the axis from some fixed point, say A. In each arm of the resonator we introduce the following formula for the coordinate system s, ξ, η:

$$\vec{R}(M) = \vec{r}(s) + \xi \vec{e}_1 + \eta \vec{e}_2, \tag{4}$$

where $\vec{R}(M)$ is the radius vector of a point M, in the k-th arm s_k varies over the interval $d_{k-1} \leq s_k \leq d_k$, and the unit vectors $\vec{e}_1(s)$ and $\vec{e}_2(s)$ are rotated about the normal $\vec{n}$ and binormal $\vec{b}$ through an angle $\vartheta = \vartheta(s)$:

$$\vec{e}_1 = \vec{n} \cos\vartheta - \vec{b} \sin\vartheta,$$

$$\vec{e}_2 = \vec{n} \sin\vartheta + \vec{b} \cos\vartheta,$$

where

$$\vartheta(s_k) = \int_{d_{k-1}}^{s_k} \varkappa(s)\,ds + \vartheta(d_{k-1}); \quad d_N \equiv d_0,$$

and $\varkappa(s)$ is the torsion (of the axis).

The Lamé coefficients of the orthogonal coordinate system constructed above are equal to

$$H_s = 1 - k(s)(\xi \cos\vartheta + \eta \sin\vartheta), \quad H_\xi = H_\eta = 1, \tag{5}$$

[$k(s)$ is the curvature of the axis].

The specification of $\vartheta(d_{k-1})\ (k = 1, 2, \ldots, N)$ determines the relation between $|s, \eta, \xi|$ and the Cartesian coordinates $|\zeta_{1i}, \zeta_{2i}, \zeta_{3i}|$. The equation for the axis of the k-th arm has the form $\xi_k = 0$, $\eta_k = 0$ in our coordinate system, and the reflection law is

$$\left.\frac{\partial s_k}{\partial \zeta_{1k}}\right|_{\substack{\zeta_{ik}=0 \\ i=1,2,3}} = \left.\frac{\partial s_{k+1}}{\partial \zeta_{1k}}\right|_{\substack{\zeta_{ik}=0 \\ i=1,2,3}}; \quad \left.\frac{\partial s_k}{\partial \zeta_{2k}}\right|_{\substack{\zeta_{ik}=0 \\ i=1,2,3}} = \left.\frac{\partial s_{k+1}}{\partial \zeta_{2k}}\right|_{\substack{\zeta_{ik}=0 \\ i=1,2,3}} = 0. \tag{6}$$

In correspondence with the parabolic equation method (see [1, 2]) we transform in Eq. (2) to coordinates x, y according to the formulas

$$x = \sqrt{\omega_0}\,\xi, \quad y = \sqrt{\omega_0}\,\eta. \tag{7}$$

The eigenfunctions $U^{(j)}$ in the j-th arm and the eigenfrequencies ω are sought in the form of series in reciprocal powers of $\sqrt{\omega_0}$:

$$U^{(j)} = \sqrt{\frac{c(s_j, 0, 0)}{c(0, 0, 0)}} \exp\left\{ i\omega_0 \int_{d_{j-1}}^{s_j} \frac{ds}{c(s, 0, 0)} \right\} \sum_{k=0}^{\infty} V_k^{(j)} \omega_0^{-\frac{k}{2}}, \tag{8}$$

$$\omega = \omega_0 + \frac{\delta_1}{\sqrt{\omega_0}} + \frac{\delta_2}{\omega_0} + \dots , \tag{9}$$

where ω_0 is the zeroth-approximation eigenfrequency (see [2]).

We shall assume that x, y and the functions V_i together with their partial derivatives are of order 1 for $\omega \to \infty$.

Substituting (5) and (7)-(9) into Eq. (2), we gather terms of like order on ω_0:

$$\left(\Delta + \frac{\omega^2}{c^2(\vec{r})}\right) U = \omega \sqrt{\frac{c(s,0,0)}{c(0,0,0)}} \exp\left(i\omega_0 \int^s \frac{ds}{c(s,0)}\right) \sum_{k=0}^{\infty} \omega^{-\frac{k}{2}} \sum_{n=0}^{k} \mathcal{L}_n V_{k-n}, \tag{10}$$

where

$$\mathcal{L}_0 = \frac{2i}{c(s,0,0)} \frac{\partial}{\partial s} + \nabla^2 - \frac{1}{c^2(s,0,0)} (K\vec{x}, \vec{x}),$$

$$\nabla = \left(\frac{\partial}{\partial x}, \frac{\partial}{\partial y}\right); \quad \vec{x} = (x, y), \tag{11}$$

and $K(s)$ is the real symmetric matrix

$$K(s) = \left\| \begin{matrix} \frac{\partial^2 c}{\partial \xi^2} & \frac{\partial^2 c}{\partial \xi \partial \eta} \\ \frac{\partial^2 c}{\partial \xi \partial \eta} & \frac{\partial^2 c}{\partial \eta^2} \end{matrix} \right\|_{s,0,0} \cdot \frac{1}{C(s,0,0)} . \tag{12}$$

In Eqs. (10) and (11) we have dropped the arm number subscript, as these equations hold in every arm of the resonator. The operators $\mathcal{L}_n$ contain the following terms:

$$P_{1n}(x,y,s)\frac{\partial^2}{\partial s^2} + P_{2n}(x,y,s)\frac{\partial}{\partial s} + P_{3n}(x,y,s)\frac{\partial}{\partial x} +$$

$$+ P_{4n}(x,y,s)\frac{\partial}{\partial y} + P_{5n}(x,y,s)\nabla^2 + \left(2\delta_n + \sum_{j=1}^{n-3} \delta_{n-2-j}\delta_j\right)\frac{1}{C^2(s,0,0)}, \quad \delta_0 = 0,$$

where $P_{in}(x,y,s)$ are homogeneous polynomials in x and y with coefficients depending on s.

In Eq. (10) we set terms with like powers of ω_0 equal to zero to obtain the following recursive system of equations:

$$\left.\begin{matrix} \mathcal{L}_0 V_0 = 0 \\ \mathcal{L}_1 V_0 + \mathcal{L}_0 V_1 = 0 \\ \dots\dots\dots\dots \\ \mathcal{L}_n V_0 + \dots + \mathcal{L}_0 V_n = 0 \\ \dots\dots\dots\dots \end{matrix}\right\}. \tag{12}$$

Next we consider the boundary conditions (3). The solutions formulated in each arm of the resonator are matched at the mirrors:

$$\left. \overset{(j)}{U} + \overset{(j+1)}{U} \right|_{s_j} = 0 \qquad (j = 1, 2, \dots, N), \tag{13}$$

and the eigenfrequencies are determined from the periodicity condition $\overset{(N+1)}{U} \equiv \overset{(1)}{U}$.

In order to satisfy the boundary conditions (13) it is convenient to write the functions $U^{(j)}$ in the single Cartesian coordinate system ξ_{1j}, ξ_{2j}, ξ_{3j} intrinsically attached to the surface S_j of the j-th mirror. We represent the expression (8) for the functions $U^{(j)}$ in the form $U^{(j)}=\sum_{k=0}^{\infty} U_k^{(j)}$ and write each component $U_k^{(j)}$ in these Cartesian coordinates. Using the equation (1) for the mirror surface, we decompose each component $U_k^{(j)}$ at S_j into a series in reciprocal powers of $\sqrt{\omega_0}$. Substituting these decompositions into Eqs. (8) and (13), from the boundary conditions we obtain, setting the coefficients of like powers of ω_0 equal to zero, a recursive system of equations, which must be satisfied by the functions $U_k^{(j)}$ and $U_k^{(j+1)}$ for $s = d_j$ identically in ξ_{1j} and ξ_{2j}.

In [2] the zeroth-approximation solution has been formulated, i.e., solutions have been found for the equation

$$\mathcal{L}_0^{(j)} V_0^{(j)} = 0, \tag{14}$$

satisfying the following condition at the mirrors:

$$U^{(j)} + U^{(j+1)} \Big|_{S_j} = O(\omega^{-\frac{1}{2}}) \tag{14'}$$

and the following condition of concentration near the resonator axis:

$$\max |V(x,y,s)| \to 0,$$
$$\sqrt{x^2+y^2} \to \infty .$$

The formulated solution satisfies Eq. (14) and conditions (14') and determines the fundamental harmonic. The higher harmonics were formulated by means of operators Λ_j and $\overset{+}{\Lambda}_i$ commuting with $\mathcal{L}_0$ and defined according to the formulas

$$\Lambda_j = (\vec{X}_j(s), \nabla) - \frac{i}{c(s,0)}\left(\frac{d\vec{X}_j(s)}{ds}, \vec{x}\right),$$
$$\overset{+}{\Lambda}_j = (\vec{X}_j^*(s), \nabla) - \frac{i}{c(s,0)}\left(\frac{d\vec{X}_j^*(s)}{ds}, \vec{x}\right), \tag{15}$$

where $\vec{X}_j$ and $\vec{X}_j^*$ are solutions of the linearized system of Euler equations for the Format functional, satisfying the conditions

$$\vec{X}_j(s+\mathcal{L}) = e^{i\varphi_j}\vec{X}_j(s),$$
$$\vec{X}_j^*(s+\mathcal{L}) = e^{-i\varphi_j}\vec{X}_j^*(s),$$

where $\mathcal{L}$ is the length of the resonator axis and φ_j are the so-called Floquet exponents (see [2]).

The following commutation relations hold for the operators Λ_j and $\overset{+}{\Lambda}_j$:

$$\Lambda_i \Lambda_j - \Lambda_j \Lambda_i = 0,$$
$$\overset{+}{\Lambda}_i \overset{+}{\Lambda}_j - \overset{+}{\Lambda}_j \overset{+}{\Lambda}_i = 0,$$

$$\Lambda_i \Lambda_j^+ - \Lambda_j^+ \Lambda_i = -\frac{\delta_{ij}}{c(s,0)}, \quad \delta_{ij} = \begin{cases} 0, & i \neq j \\ 1, & i = j \end{cases}.$$

Applied to v (where v is the fundamental harmonic of the resonator), the operators Λ_j^+ give us the higher harmonics:

$$V_{o[m]} = (\Lambda_1^+)^{m_1} (\Lambda_2^+)^{m_2} v = \sum_{\ell_1 + \ell_2 = m_1 + m_2} a_{\ell_1 \ell_2}(s) x^{\ell_1} y^{\ell_2} v. \tag{16}$$

The functions $\Lambda_1^{+\lambda_1} \Lambda_2^{+\lambda_2} v$ are orthogonal, and $\Lambda_j v = 0$.

The eigenfrequencies $\omega_{o[m]}$ are described by the formula

$$\omega_{o[m]} \int_0^{\mathcal{L}} \frac{ds}{c(s,0)} = 2\pi p + N\pi + (m_1 + \tfrac{1}{2})\varphi_1 + (m_2 + \tfrac{1}{2})\varphi_2, \tag{17}$$

$$m_1, m_2 = 0, 1, 2, \ldots.$$

We are now ready to formulate the higher approximations for the fundamental harmonic.

We seek the solutions to the recursive system of equations (12) in the form

$$V_i = \sum_{\lambda_1 + \lambda_2 \leq 3i} a^{(i)}_{\lambda_1 \lambda_2}(s) (\Lambda_1^+)^{\lambda_1} (\Lambda_2^+)^{\lambda_2} v. \tag{18}$$

The operators of multiplication by x and y and differentiation with respect to x and y, applied to functions of the given class, can be expressed by solving the system (15) by means of a linear combination of Λ_j and Λ_j^+. Using the explicit form of the operator $\frac{\partial}{\partial s}$, we can write the differentiation operator $\mathcal{L}_0$ in the form

$$\frac{\partial}{\partial s} = \frac{c(s,0,0)}{2i} \left\{ \mathcal{L}_0 - \sum_{i,j=1}^{2} \left\{ \alpha_{ij}(s) \Lambda_i \Lambda_j + \beta_{ij} \Lambda_j^+ \Lambda_i + \varrho_{ij} \Lambda_i^+ \Lambda_j^+ \right\} \right\}.$$

With the foregoing in mind, we obtain the following expression for $\mathcal{L}_0 V_n$:

$$\mathcal{L}_0 V_n = \sum_{\lambda_1 + \lambda_2 \leq 3n} f^{(n)}_{\lambda_1 \lambda_2}(s) \Lambda_1^{+\lambda_1} \Lambda_2^{+\lambda_2} v - \frac{2\delta_n}{c^2(s,0,0)} v. \tag{19}$$

Inserting (18) into Eq. (19), we transform to the following equations for $a_{\lambda_1 \lambda_2}(s)$:

$$\frac{2i}{c(s,0,0)} \frac{d}{ds} a^{(n)}_{\lambda_1 \lambda_2}(s) = f^{(n)}_{\lambda_1 \lambda_2}(s) \tag{20}$$

and

$$\frac{2i}{c(s,0,0)} \frac{d}{ds} a^{(n)}_{00}(s) = f^{(n)}_{00} - \frac{2\delta_n}{c^2(s,0,0)}. \tag{21}$$

We need to write the boundary conditions for $U_n^{(j)}$ by means of Λ_k^+ in a workable form. We decompose each function $U_n^{(j)}$ at the mirror surface S_j into a Taylor series in the vicinity of the point d_j. We obtain series in reciprocal powers of $\sqrt{\omega_0}$, the terms of which are polynomials in x and y. Substituting these decompositions into (13) and redecomposing the polynomials in x and y on the operator polynomials, after suitable computations we arrive at the following conditions for the coefficients $a^{(n)}_{\lambda_1 \lambda_2}(s)$ at the points $d_1, d_2, \ldots, d_N$:

$$a^{(n)}_{\lambda_1\lambda_2}(d_\kappa+0)-a^{(n)}_{\lambda_1\lambda_2}(d_\kappa-0)=\gamma^{(n)}_{\lambda_1\lambda_2}(d_\kappa), \tag{22}$$

where $\gamma^{(n)}_{\lambda_1\lambda_2}(d_\kappa)$ are known quantities depending in particular on the geometry of mirror S_κ, i.e., on the coefficients $g^{(\kappa)}_{mn}$ on Eq. (1).

In Eq. (20) for $a^{(n)}_{\lambda_1\lambda_2}(s)$ we transform to the periodic functions $\hat{a}^{(n)}_{\lambda_1\lambda_2}(s)$, separating out the periodic parts from $a^{(n)}_{\lambda_1\lambda_2}(s)$ and $f^{(n)}_{\lambda_1\lambda_2}(s)$:

$$\begin{gathered}a^{(n)}_{\lambda_1\lambda_2}(s)=\exp\left(is\frac{\varphi_1\lambda_1+\varphi_2\lambda_2}{\mathcal{L}}\right)\hat{a}^{(n)}_{\lambda_1\lambda_2}(s),\\ f^{(n)}_{\lambda_1\lambda_2}(s)=\exp\left\{i\frac{s}{\mathcal{L}}(\varphi_1\lambda_1+\varphi_2\lambda_2)\right\}\tilde{f}^{(n)}_{\lambda_1\lambda_2}(s).\end{gathered} \tag{23}$$

From (20) and (22) we obtain the following problems for $\hat{a}^{(n)}_{\lambda_1\lambda_2}(s)$:

$$\begin{gathered}\frac{2i}{c(s,0,0)}\frac{d}{ds}\hat{a}^{(n)}_{\lambda_1\lambda_2}(s)+\frac{2i}{c(s,0,0)}E_{\lambda_1\lambda_2}\hat{a}^{(n)}_{\lambda_1\lambda_2}(s)=\tilde{f}^{(n)}_{\lambda_1\lambda_2},\\ \hat{a}^{(n)}_{\lambda_1\lambda_2}(d_\kappa+0)-\hat{a}^{(n)}_{\lambda_1\lambda_2}(d_\kappa-0)=\gamma(d_\kappa)e^{-E_{\lambda_1\lambda_2}d_\kappa},\\ \hat{a}^{(n)}_{\lambda_1\lambda_2}(s+\mathcal{L})=\hat{a}^{(n)}_{\lambda_1\lambda_2}(s),\end{gathered} \tag{24}$$

where $E_{\lambda_1\lambda_2}=i\,\dfrac{\varphi_1\lambda_1+\varphi_2\lambda_2}{\mathcal{L}}$.

If the function $e^{-E_{\lambda_1\lambda_2}s}$ is aperiodic, problem (24) is always solvable. In the case of periodic $e^{-E_{\lambda_1\lambda_2}s}$ the homogeneous problem has a nontrivial solution, and for (24) to be solvable it is necessary that the following orthogonality condition be fulfilled for $\tilde{f}^{(n)}_{\lambda_1\lambda_2}(s)$:

$$\frac{1}{2i}\int_0^{\mathcal{L}}\tilde{f}^{(n)}_{\lambda_1\lambda_2}(s)c(s,0,0)ds=\sum_{\kappa=1}^{N}\gamma(d_\kappa). \tag{25}$$

If it is not fulfilled, then a periodic solution does not exist for $\hat{a}^{(n)}_{\lambda_1\lambda_2}(s)$. In our notation the periodicity condition is equivalent to linear dependence of the numbers φ_1,φ_2, and 2π over the ring of integers. Equation (24) is a linear inhomogeneous equation. Its solution takes the form

$$\begin{gathered}\hat{a}^{(n)}_{\lambda_1\lambda_2}(s)=e^{-E_{\lambda_1\lambda_2}s}\left(\frac{1}{2i}\int_0^s\tilde{f}^{(n)}_{\lambda_1\lambda_2}(s)e^{E_{\lambda_1\lambda_2}s}c(s,0)ds+C\right),\\ C\equiv\text{const}.\end{gathered} \tag{26}$$

Each arm of the resonator is endowed with its own value of C.

We substitute (26) into the boundary conditions, obtaining the following relation between the C_j:

$$\begin{gathered}\gamma(d_1)=C_2-C_1,\\ \cdots\cdots\cdots\\ \gamma(d_{N-1})=C_N-C_{N-1},\\ \gamma(d_N)=C_{N+1}-C_N,\end{gathered}$$

$$c_1 = e^{-E_{\lambda_1\lambda_2}\mathcal{L}} \left(\frac{1}{2i} \int_0^{\mathcal{L}} \tilde{f}^{(n)}_{\lambda_1\lambda_2}(s) c(s,0) e^{E_{\lambda_1\lambda_2}s} c(s + C_{N+1} \right).$$

From this we can determine all the c_j in the case of aperiodic $e^{-E_{\lambda_1\lambda_2}s}$.

Consider the equation

$$\frac{2i}{c(s,0,0)} \frac{d}{ds} a^{(n)}_{00}(s) = f^{(n)}_{00} - \frac{2\delta_n}{c^2(s,0,0)} \tag{27}$$

with the following condition at the points d_κ:

$$a^{(n)}_{00}(d_\kappa + 0) - a^{(n)}_{00}(d_\kappa - 0) = \gamma^{(n)}_{00}(d_\kappa),$$

$$a^{(n)}_{00}(s + \mathcal{L}) = a^{(n)}_{00}(s).$$

From the condition for solvability of Eq. (27) we determine the correction to the eigenfrequency $\omega_{0[0]}$:

$$\delta_{n[0]} = \frac{\int_0^{\mathcal{L}} f^{(n)}_{00} c(s,0) ds + 2i \sum_{\kappa=1}^{N} \gamma^{(n)}_{00}(d_\kappa)}{2 \int_0^{\mathcal{L}} c^{-1}(s,0) ds}. \tag{28}$$

Proceeding by analogy with [3], we can show that $f^{(n)}_{00}$'s and $\gamma^{(n)}_{00}(d_\kappa)$ are nonzero quantities only in the case of even n.

Consequently, the eigenfrequency is

$$\omega = \omega_0 + \frac{\delta_2}{\omega_0} + \frac{\delta_4}{\omega_0^2} + \dots, \tag{29}$$

and the eigenfunction is

$$V = \sum_{i=0}^{\infty} \frac{V_i}{(\sqrt{\omega_0})^i}.$$

The formulations carried out for the fundamental harmonic without singular variations carries over to the higher harmonics. If we have

$$V_{0[m]} = (\Lambda_1^+)^{m_1} (\Lambda_2^+)^{m_2} v$$

for the zeroth approximation, then we seek the k-th approximation in the form

$$V_{\kappa[m]} = \sum_{\lambda_1 + \lambda_2 \le m_1 + m_2 + 3\kappa} a^{(\kappa, m_1, m_2)}_{\lambda_1 \lambda_2} (\Lambda_1^+)^{\lambda_1} (\Lambda_2^+)^{\lambda_2} v. \tag{30}$$

As for the fundamental harmonic, the problem reduces to the solution of linear differential equations for the coefficients $a^{(\kappa; m_1 m_2)}_{\lambda_1 \lambda_2}$. We determine the corrections to the eigenfrequency [see (17)] from the solvability condition for $a^{(\kappa, m_1 m_2)}_{m_1 m_2}(s)$.

In conclusion the author wishes to thank V. M. Babich for stating the problem and aiding in its solution.

LITERATURE CITED

1. Babich, V. M., Eigenfunctions concentrated in a neighborhood of a closed geodesic, Seminars in Mathematics, Vol. 9: Mathematical Problems in Wave Propagation Theory, Consultants Bureau, New York (1970), pp. 7-26.
2. Popov, M. M., Eigenmodes of a multimirror resonator, Vest. Leningrad. Univ., (in press).
3. Pyshkina, M. F., Asymptotic behavior of eigenfunctions of the Helmholtz equation concentrated near a closed geodesic, this volume, p. 88.

SHADOW ZONE AT A FIRST-ORDER INTERFACE BETWEEN INHOMOGENEOUS MEDIA

N. V. Tsepelev

§ 1. Statement of the Problem

In a Cartesian coordinate system (x, z) let there be given two half-spaces: $z < 0$ $(\nu = 1)$ and $z > 0$ $(\nu = 2)$, in which the propagation of oscillations is described by wave equations and the velocities are such that $v_1 = v_1(x)$ and $v_2 = v_2(x, z)$. We make the following assumptions with regard to the functions $v_1(x)$ and $v_2(x, z)$: 1) They are sufficiently smooth positive functions defined and bounded for (respectively) $|x| < \infty$ and $|x| < \infty$, $z > 0$; 2) $v_1(x) = v_2(x, 0) = v(x)$; 3) $\frac{\partial v_2(x,z)}{\partial z} < 0$; 4) they vary sufficiently slowly as a function of x. We also require that condition (1.3) introduced below be satisfied.

It is required to investigate for high frequencies the steady-state wave field $u_\nu = U_\nu(x, z)e^{-i\omega t}$ from a point source situated in the half-space $z > 0$ at a point M_0 with coordinates $x_0 = 0$ and $z_0 = H$.

The functions $U_\nu(x, z)$ satisfy the equations

$$\Delta U_1 + \frac{\omega^2}{v_1^2(x)} U_1 = 0 \qquad (z < 0),$$

$$\Delta U_2 + \frac{\omega^2}{v_2^2(x,z)} U_2 = -\delta(x)\delta(z - H) \qquad (z > 0), \tag{1.1}$$

where δ is the Dirac delta function. We require at $z = 0$ that the functions U_ν satisfy the conjugacy conditions

$$U_1\Big|_{z=0} = U_2\Big|_{z=0}, \quad \frac{\partial U_1}{\partial z}\Big|_{z=0} = \frac{\partial U_2}{\partial z}\Big|_{z=0}, \tag{1.2}$$

and that as $|x^2 + z^2| \to \infty$ the principle of limiting absorption holds.

Under the stated assumptions with regard to the functions v_ν for $z > 0$, near the boundary $z = 0$ a geometric shadow zone is formed (Fig. 1). Let $z = f(x, \alpha)$ be the equation for a ray emanating from the source, and let α be the angle between the tangent to the ray at the source and the z axis. We propose to find an $\alpha = \alpha_0$ such that the corresponding ray touches the x axis at some point a and returns into the half-space $z > 0$. The given ray to the right of a joins with the x axis to form the boundaries of the geometric shadow zone. If the inequality

$$\frac{\partial v_2(x,z)}{\partial x} f'(x, \alpha) - \frac{\partial v_2(x,z)}{\partial z} \geqslant 0 \tag{1.3}$$

can be satisfied (where the rays have curvature of constant sign and there are no inflection points) and caustics are absent everywhere, then the propagation pattern of the rays illustrated in Fig. 1 is preserved throughout the entire half-space $z > 0$.

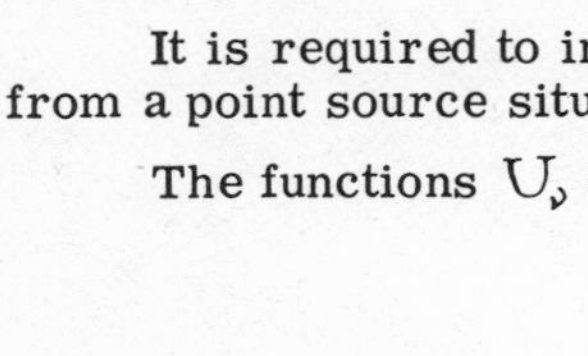

Fig. 1.

Whereas in the illuminated zone the field can be found by the methods of geometrical optics (by the ray method, for example), it is necessary in the shadow zone to employ special techniques. In the present article, which is concerned with the shadow zone, we use the parabolic equation method.

§ 2. Formulation of the Solution

To the right of the point a (Fig. 1) we consider the layer immediately continguous to the x axis with a thickness assumed of the order $O(\omega^{-2/3})$, and we carry out the ensuing analyses for z satisfying the inequality

$$|z| < A\omega^{-2/3}, \quad A = \text{const.} \tag{2.1}$$

In this layer we introduce the attenuation function

$$V_\nu = U_\nu e^{-i\omega \int_0^x \frac{dx}{v(x)} - \omega^{1/3} h(x)}, \tag{2.2}$$

where $h(x)$ is arbitrary for the time being, and we require that $U_\nu(x, z)$ satisfy the homogeneous system of equations with respect to (1.1).

If we introduce the new independent variable

$$\zeta = \omega^{2/3} \varphi(x) z, \tag{2.3}$$

where $\varphi(x)$ is to be defined below, and decompose the difference $v_2^{-2}(x, z) - v_1^{-2}(x)$ into a power series in z, then, stopping with terms of order $O(\omega^{-2/3})$, we obtain the following equations for $V_\nu(x, z)$:

$$\left.\begin{aligned}
&\frac{\partial^2 V_1}{\partial \zeta^2} + \frac{2i h'(x)}{v(x)\varphi^2(x)} V_1 + \frac{2i}{\omega^{1/3} v(x)\varphi^2(x)}\left[\frac{\partial V_1}{\partial x} + \zeta\frac{\varphi'(x)}{\varphi(x)}\frac{\partial V_1}{\partial \zeta} - \frac{1}{2}\frac{v'(x)}{v(x)} V_1\right]\left[1 + O(\omega^{-1/3})\right] = 0,\\
&\frac{\partial^2 V_2}{\partial \zeta^2} + \left[\frac{2i h'(x)}{v(x)\varphi^2(x)} + \frac{2\varepsilon(x)\zeta}{v^2(x)\varphi^3(x)}\right] V_2 + \frac{2i}{\omega^{1/3} v(x)\varphi^2(x)}\left[\frac{\partial V_2}{\partial x} + \zeta\frac{\varphi'(x)}{\varphi(x)}\frac{\partial V_2}{\partial \zeta} - \frac{1}{2}\frac{v'(x)}{v(x)} V_2\right]\left[1 + O(\omega^{-1/3})\right] = 0.
\end{aligned}\right\} \tag{2.4}$$

Here we have assumed

$$\varepsilon(x) = -\frac{1}{v(x)} \left.\frac{\partial v_2(x, z)}{\partial z}\right|_{z=0} > 0. \tag{2.5}$$

The functions $h(x)$ and $\varphi(x)$ are chosen so as to make the coefficients in the principal terms of the system (2.4) independent of x. This requires that

$$\varphi(x) = \left[\frac{2\varepsilon(x)}{v^2(x)}\right]^{1/3}, \tag{2.6}$$

$$h(x) = \frac{ip}{2}\int_0^x v(x)\varphi^2(x)\,dx, \tag{2.7}$$

where p is an arbitrary constant.

We shall assume that $\varepsilon(x)$ is bounded, whereupon this same property is invested in the function $\varphi(x)$. Consequently, ζ also remains bounded in the given layer as $\omega \to \infty$.

Using (2.6) and (2.7), we obtain

$$\left.\begin{aligned}&\frac{\partial^2 V_1}{\partial s^2}-pV_1-\frac{p}{\omega^{1/3}h'(x)}\left[\frac{\partial V_1}{\partial x}+s\frac{\varphi'(x)}{\varphi(x)}\frac{\partial V_1}{\partial s}-\frac{1}{2}\frac{v'(x)}{v(x)}V_1\right]\left[1+O(\omega^{-1/3})\right]=0,\\&\frac{\partial^2 V_2}{\partial s^2}-(p-s)V_2-\frac{p}{\omega^{1/3}h'(x)}\left[\frac{\partial V_2}{\partial x}+s\frac{\varphi'(x)}{\varphi(x)}\frac{\partial V_2}{\partial s}-\frac{1}{2}\frac{v'(x)}{v(x)}V_2\right]\left[1+O(\omega^{-1/3})\right]=0.\end{aligned}\right\}\tag{2.8}$$

In order to separate the variables in terms of order $O(\omega^{-1/3})$ it suffices to perform the substitution

$$V_\nu=W_\nu\exp\left\{\omega^{-1/3}s^2\frac{p}{2\varphi^2(x)}\int_0^x\frac{[\varphi'(x)]^2}{h'(x)}dx\right\}.\tag{2.9}$$

It turns out that W_ν must satisfy the equations

$$\begin{aligned}&\frac{\partial^2 W_1}{\partial s^2}-pW_1-\frac{p}{\omega^{1/3}h'(x)}\left\{\frac{\partial W_1}{\partial x}-\frac{1}{2}\left[\frac{\varphi''(x)}{\varphi'(x)}-2\frac{\varphi'(x)}{\varphi(x)}\right]W_1\right\}\left[1+O(\omega^{-1/3})\right]=0,\\&\frac{\partial^2 W_2}{\partial s^2}-(p-s)W_2-\frac{p}{\omega^{1/3}h'(x)}\left\{\frac{\partial W_2}{\partial x}-\frac{1}{2}\left[\frac{\varphi''(x)}{\varphi'(x)}-2\frac{\varphi'(x)}{\varphi(x)}\right]W_2\right\}\left[1+O(\omega^{-1/3})\right]=0,\end{aligned}\tag{2.10}$$

whose solutions are easily found:

$$\begin{aligned}&W_1=\frac{\sqrt{\varphi'(x)}}{\varphi(x)}e^{\pm\sqrt{p}s}\left[1+O(\omega^{-1/3})\right],\\&W_2=\frac{\sqrt{\varphi'(x)}}{\varphi(x)}w_n(p-s)\left[1+O(\omega^{-1/3})\right].\end{aligned}\tag{2.11}$$

Here w_n is an Airy function. In the complex plane of p it is required to make a cut and fix a branch of the function $\sqrt{p}$. We make the cut from 0 to $-\infty$ and let $\sqrt{p}>0$ for $p>0$.

Using expressions (2.2), (2.9), and (2.11), we can represent the solution of the initial homogeneous equations as follows:

$$\begin{aligned}&U_1=M\frac{\sqrt{\varphi'(x)}}{\varphi(x)}e^{\pm\sqrt{p}s+i\omega\int_0^x\frac{dx}{v(x)}+\frac{ip}{2}\omega^{1/3}\int_0^x v(x)\varphi^2(x)dx}\left[1+O(\omega^{-1/3})\right],\\&U_2=N\frac{\sqrt{\varphi'}}{\varphi(x)}w_n(p-s)e^{i\omega\int_0^x\frac{dx}{v(x)}+\frac{ip}{2}\omega^{1/3}\int_0^x v(x)\varphi^2(x)dx}\left[1+O(\omega^{-1/3})\right],\end{aligned}\tag{2.12}$$

where M and N are arbitrary constants.

The case when the velocities are independent of x differs from the case just considered to the extent that $\varphi=\text{const}$ and the factor $\frac{\sqrt{\varphi'(x)}}{\varphi(x)}$ has to be replaced by unity.

We require that the solutions (2.12) satisfy the limiting absorption principle and the boundary conditions (1.2). Let us assume that the frequency ω has an imaginary part with a small positive coefficient or, equivalently, that the medium has small damping. Then in order for the solution to tend to zero as $x^2+z^2\to\infty$ it is necessary in U_2 to adopt the first Airy function, i.e., $w_1(p-s)$. In the upper half-space $(z<0)$, on the other hand, the field vanishes if the solution $e^{\sqrt{p}s}$ is chosen with

$$0 < \arg p < \pi. \tag{2.13}$$

However, the limiting absorption principle is also satisfied by the functions $e^{\pm\sqrt{p}s}$ with negative p (on the upper and lower rims of the cut of the p plane). This means that $U_\nu(x,z)$ can be represented by the sum

$$U_\nu(x,z) = \left[\alpha U_\nu^{(0)}(x,z) + \beta U_\nu^{(1)}(x,z)\right]\left[1+O(\omega^{-1/3})\right]. \tag{2.14}$$

Here we have set

$$\left.\begin{aligned} U_1^{(0)} &= A(p)e^{\sqrt{p}s}\Phi(x,p), \\ U_2^{(0)} &= B(p)w_1(p-s)\Phi(x,p); \end{aligned}\right\} \tag{2.15}$$

$$\left.\begin{aligned} U_1^{(1)} &= \left[C(q)e^{is\sqrt{q}} + D(q)e^{-is\sqrt{q}}\right]\Phi(x,-q), \\ U_2^{(1)} &= \mathscr{E}(q)w_1(-q-s)\Phi(x,-q), \end{aligned}\right\} \tag{2.16}$$

where

$$\Phi(x,p) = \frac{\sqrt{\varphi'(x)}}{\varphi(x)} e^{i\omega\int_0^x \frac{dx}{v(x)} + \frac{ip}{2}\omega^{1/3}\int_0^x v(x)\varphi^2(x)dx}, \tag{2.17}$$

and α, β are arbitrary constants. The need for their introduction is stipulated by the following consideration. Inasmuch as we have formulated solutions $U_\nu(x,z)$ to terms of order $O(\omega^{-1/3})$, it can happen that one of the terms in (2.14) will be of order $O(\omega^{-1/3})$ relative to the other. In this event it is necessary to set the corresponding coefficient (α or β) equal to zero and the other equal to one. But if they are of the same order of smallness on ω, then α and β must be assumed equal. It will be convenient below to have

$$\alpha + \beta = 1. \tag{2.18}$$

The unknown coefficients in (2.15) and (2.16) can be determined from the boundary conditions independently of one another. Let us first consider (2.15). Substituting $U_\nu^{(0)}$ into (1.2) for $A(p)$ and $B(p)$, we obtain the system

$$\left.\begin{aligned} A(p) - B(p)w_1(p) &= 0, \\ \sqrt{p}A(p) + B(p)w_1'(p) &= 0. \end{aligned}\right\} \tag{2.19}$$

Inasmuch as the system is homogeneous, in order for it to be solvable it is necessary that the parameter p satisfy the equation

$$\sqrt{p}\,w_1(p) + w_1'(p) = 0, \tag{2.20}$$

the roots of which are given approximately by the formula

$$p_n \approx \left(\frac{3n\pi}{2}\right)^{2/3} e^{i\left(\frac{\pi}{3} + \frac{\ln 12n\pi}{3n\pi}\right)}, \tag{2.21}$$

the accuracy of the formula increasing with n.

Note that all the roots satisfy inequality (2.13). It follows from (2.19) that

$$A(p_n) = B(p_n)w_1(p_n) \tag{2.22}$$

and the solution (2.15) assumes the form

$$\left.\begin{aligned} U_1^{(0)} &= B(p_n)w_1(p_n)e^{5\sqrt{p_n}}\Phi(x, p_n), \\ U_2^{(0)} &= B(p_n)w_1(p_n - 5)\Phi(x, p_n). \end{aligned}\right\} \tag{2.23}$$

If we insert the solution (2.16) into the boundary conditions (1.2), it turns out that the coefficients satisfy the system

$$\left.\begin{aligned} C(q) - \mathcal{E}(q)w_1(-q) &= -\mathcal{D}(q), \\ i\sqrt{q}\,C(q) + \mathcal{E}(q)w_1'(-q) &= i\sqrt{q}\,\mathcal{D}(q). \end{aligned}\right\} \tag{2.24}$$

Since (2.20) does not have any negative roots, the system (2.24) is solvable for any q. We introduce the notation

$$\left.\begin{aligned} f(q) &= i\sqrt{q}\,w_1(-q) + w_1'(-q), \\ \tilde{f}(q) &= i\sqrt{q}\,w_1(-q) - w_1'(-q); \end{aligned}\right\} \tag{2.25}$$

$$M(q) = \tilde{f}(q)\mathcal{D}(q). \tag{2.26}$$

Solving (2.24) with the use of (2.25) and (2.26), we obtain

$$\left.\begin{aligned} U_1^{(1)} &= M(q)\left[\frac{e^{i5\sqrt{q}}}{f(q)} + \frac{e^{-i5\sqrt{q}}}{\tilde{f}(q)}\right]\Phi(x, -q), \\ U_2^{(1)} &= M(q)\,\frac{2i\sqrt{q}\,w_1(-q-5)}{f(q)\tilde{f}(q)}\,\Phi(x, -q). \end{aligned}\right\} \tag{2.27}$$

§ 3. Determination of the Factors $B(p_n)$ and $M(q)$

In order to find the factors $B(p_n)$ and $M(q)$ from (2.23) and (2.27) we use the method described in [1]. It comprises the following. By means of the solution (2.14) we formulate the Green function of the problem (1.1)-(1.2) for the domain of the plane (x, z) defined by the inequalities $x > a$ and (2.1), and we find $B(p_n)$ and $M(q)$ so as to formally satisfy Eqs. (1.1) up to principal terms. This technique can only be substantiated indirectly. In problems admitting the formulation of an exact solution it yields correct results.

Let us first determine $B(P_n)$. For this we investigate the functions

$$\left.\begin{aligned} y_1(P_n, \varsigma) &= w_1(P_n)e^{\varsigma\sqrt{P_n}}, \\ y_2(P_n, \varsigma) &= w_1(P_n - \varsigma). \end{aligned}\right\} \tag{3.1}$$

Taking account of Eqs. (2.23), we use them to represent the solution satisfying conditions (1.2) for the system homogeneous with respect to (1.1) in the form

$$\left.\begin{aligned} U_1^{(0)+} &= y_1(P_n, \varsigma)\Phi(x, P_n), \\ U_2^{(0)+} &= y_2(P_n, \varsigma)\Phi(x, P_n). \end{aligned}\right\} \tag{3.2}$$

The plus sign on $U_\nu^{(0)+}$ denotes that the given expressions describe oscillations propagating in the direction of increasing x $(x > a)$. Formally changing the sign in front of x in $\Phi(x, P_n)$, we obtain expressions for $U_\nu^{(0)-}$, describing the field for negative x $(x < -\tilde{a})$. It is reasonable to assume that the function $B(P_n)$ we seek depends not only on the parameter P_n, but also on the position of the source, i.e., that $B(P_n) = B(P_n, 0, \varsigma_0)$, where

$$\varsigma_0 = \omega^{2/3}\varphi(0)H. \tag{3.3}$$

We assign $B(P_n, 0, \varsigma_0)$ the + or - sign in compliance with the foregoing.

Inasmuch as the stated problem is not only satisfied by $U_\nu^{(0)+}$ and $U_\nu^{(0)-}$, but also by their sums on all the roots P_n, we can write

$$U_\nu^{(0)} = \sum_{n=1}^{\infty} \begin{cases} B^+(P_n, 0, \varsigma_0)U_\nu^{(0)+}(P_n, x, \varsigma), \\ B^-(P_n, 0, \varsigma_0)U_\nu^{(0)-}(P_n, x, \varsigma), \end{cases} \tag{3.4}$$

which is also a solution of our problem.

In order to satisfy the reciprocity principle, we need to put

$$B^{\pm}(P_n, 0, \varsigma_0) = N(P_n)U_\nu^{(0)\mp}(P_n, 0, \varsigma_0), \tag{3.5}$$

where $N(P_n)$ is independent of the position of the source and the point of observation. Therefore, using the form of $U_\nu^{(0)\pm}$, we obtain

$$U_\nu^{(0)} = \frac{\sqrt{\varphi'(0)}}{\varphi(0)} \sum_{n=1}^{\infty} N(P_n) y_2(P_n, \varsigma_0) y_\nu(P_n, \varsigma)\Phi(|x|, P_n). \tag{3.6}$$

If we consider the functions

$$\left.\begin{aligned} \tilde{y}_1(q, \varsigma) &= \frac{e^{i\varsigma\sqrt{q}}}{f(q)} + \frac{e^{-i\varsigma\sqrt{q}}}{\tilde{f}(q)}, \\ \tilde{y}_2(q, \varsigma) &= \frac{2i\sqrt{q}\, w_1(-q-\varsigma)}{f(q)\tilde{f}(q)}, \end{aligned}\right\} \tag{3.7}$$

then, pursuing analogous arguments, we arrive at the following expressions for $U_\nu^{(1)}$:

$$U_\nu^{(1)} = \frac{\sqrt{\varphi'(0)}}{\varphi(0)} \int_0^\infty Q(q)\tilde{y}_2(q,s_0)\tilde{y}_\nu(q,s)\Phi(|x|,-q)\,dq. \tag{3.8}$$

We shall say more about the convergence of the integral (3.8) in § 4, merely noting for the time being that all the arguments remain in effect if the contour is deformed into the upper half-plane.

We have thus reduced the problem to the determination of the unknown factors $N(P_n)$ and $Q(q)$ from (3.6) and (3.8).

We now set down a property of the functions (3.1) and (3.7) that will be important to us later on. It is readily perceived that they are solutions of the differential equation

$$y_\nu'' + s\varepsilon(s)y_\nu = \mathfrak{s}y_\nu, \tag{3.9}$$

where

$$\varepsilon(s) = \begin{cases} 0, & s < 0, \\ 1, & s > 0, \end{cases}$$

and the parameter $\mathfrak{s}$ assumes values P_n or $-q$. They meet the following conditions in this case:

$$y_1\Big|_{s=0} = y_2\Big|_{s=0}; \quad \frac{\partial y_1}{\partial s}\Big|_{s=0} = \frac{\partial y_2}{\partial s}\Big|_{s=0}. \tag{3.10}$$

Inasmuch as the functions (3.1) were chosen with regard for the fact that $y_\nu \xrightarrow[s\to\pm\infty]{} 0$, we can in the usual manner by means of (3.9) and (3.10) state the following equation for them:

$$\int_{-\infty}^{\infty} y_\nu(P_k,s)y_\nu(P_r,s)\,ds = \begin{cases} 0, & k \neq r, \\ \dfrac{w_1^2(P_r)}{2\sqrt{P_r}}, & k = r. \end{cases} \tag{3.11}$$

Next we consider the solutions (3.7). Taking account of (3.9), (3.10), and the fact that $\tilde{y}_2 \xrightarrow[s\to\infty]{} 0$, we obtain

$$\int_{-A}^{\infty} \tilde{y}_\nu(q_1,s)\tilde{y}_\nu(q_2,s)\,ds = \frac{[\tilde{y}_1(q_1,s)\tilde{y}_1'(q_2,s) - \tilde{y}_1'(q_1,s)\tilde{y}_1(q_2,s)]\big|_{s=-A}}{q_2 - q_1}, \tag{3.12}$$

where $A > 0$ is arbitrary. Multiplying the given relation by any continuous function $\Phi(q_2)$ and integrating over q_2 from 0 to ∞, we have for $A \to \infty$

$$\int_0^\infty \Phi(q_2)\,dq_2 \int_{-\infty}^{\infty} \tilde{y}_\nu(q_1,s)\tilde{y}_\nu(q_2,s)\,ds = -\frac{4\sqrt{q_1}\,\pi\,\Phi(q_1)}{f(q_1)\bar{f}(q_1)}. \tag{3.13}$$

Relations (3.11) and (3.13) make it possible to find the unknown coefficients $N(P_n)$ and $Q(q)$.

We require for this that the solution (2.14) formally satisfy the inhomogeneous system of equations (1.1), writing the expression on the right-hand side as follows: $-(\alpha+\beta)\delta(x)\omega^{2/3}\varphi(0)\delta(s-s_0)$. Differentiating (2.14) and setting the principal parts of the coefficients after α and β equal to zero, we obtain

$$\frac{2i\omega}{v(0)}\left[\frac{\sqrt{\varphi'(0)}}{\varphi(0)}\right]^2 \sum_{n=1}^{\infty} N(P_n)y_2(P_n,s_0)y_\nu(P_n,s) = -\omega^{2/3}\varphi(0)\delta(s-s_0),$$

$$\frac{2i\omega}{v(0)}\left[\frac{\sqrt{\varphi'(0)}}{\varphi(0)}\right]^2 \int_0^\infty Q(q)\tilde{y}_2(q,\zeta_0)\tilde{y}_\nu(q,\zeta)\,dq = -\omega^{2/3}\varphi(0)\,\delta(\zeta-\zeta_0). \tag{3.14}$$

If we multiply the given equations by $y_\nu(p_n,\zeta)$ and $\tilde{y}_\nu(q_1,\zeta)$, respectively, and integrate over ζ from $-\infty$ to ∞ using (3.11) and (3.13), we ascertain that

$$\left.\begin{aligned} &N(p_n)\frac{2i\omega}{v(0)}\left[\frac{\sqrt{\varphi'(0)}}{\varphi(0)}\right]^2\frac{w_1'^2(p_n)}{2\sqrt{p_n}}\,y_2(p_n,\zeta_0) = -\omega^{2/3}\varphi(0)\,y_2(p_n,\zeta_0),\\ &Q(q)\frac{8\pi i\omega\sqrt{q_1}}{v(0)f(q_1)\tilde{f}(q_1)}\left[\frac{\sqrt{\varphi'(0)}}{\varphi(0)}\right]^2\tilde{y}_2(q_1,\zeta_0) = \omega^{2/3}\varphi(0)\,\tilde{y}_2(p_n,\zeta_0). \end{aligned}\right. \tag{3.15}$$

Hence

$$\left.\begin{aligned} N(p_n) &= \omega^{-1/3}\frac{v(0)\varphi^3(0)}{\varphi'(0)}\frac{2\sqrt{p_n}}{w_1'^2(p_n)},\\ Q(q) &= -\frac{i}{8\pi}\frac{\omega^{-1/3}v(0)\varphi^3(0)}{\varphi'(0)\sqrt{q}}\,f(q)\tilde{f}(q). \end{aligned}\right\} \tag{3.16}$$

Substituting the resulting expressions into (3.6) and (3.8), we finally obtain

$$\begin{aligned} U_\nu^{(0)} &= \frac{v(0)}{\omega^{1/3}}\frac{\varphi^2(0)}{\sqrt{\varphi'(0)}}\sum_{n=1}^\infty\frac{i\sqrt{p_n}}{w_1'^2(p_n)}\,y_2(p_n,\zeta_0)\,y_\nu(p_n,\zeta)\,\Phi(|x|,p_n)\left[1+O(\omega^{-1/3})\right],\\ U_\nu^{(1)} &= -\frac{i}{8\pi}\frac{v(0)}{\omega^{1/3}}\frac{\varphi^2(0)}{\sqrt{\varphi'(0)}}\int_0^\infty\frac{f(q)\tilde{f}(q)}{\sqrt{q}}\,\tilde{y}_2(q,\zeta_0)\,\tilde{y}_\nu(q,\zeta)\,\Phi(|x|,-q)\,dq\left[1+O(\omega^{-1/3})\right]. \end{aligned} \tag{3.17}$$

§ 4. Separation of the Principal Part of the Field in the Shadow Zone

Insofar as we are interested in the behavior of the field in the shadow zone created for $z>0$, we shall only investigate $U_2(x,z)$ below. We first consider the component $U_2^{(1)}$ for $x>a$. It may be represented by the following contour integral:

$$U_2^{(1)} = \frac{v(0)}{4\pi\omega^{1/3}}\frac{\varphi^2(0)}{\sqrt{\varphi'(0)}}\int_{(\lambda)}\frac{w_1(s-\zeta_0)\,w_1(s-\zeta)\,\Phi(x,s)\,ds}{w_1(s)\left[\sqrt{s}\,w_1(s)+w_1'(s)\right]}\left[1+O(\omega^{-1/3})\right], \tag{4.1}$$

where the contour (λ) encloses the cut drawn in the s plane from the origin in the direction of $e^{i\psi}$, where $\frac{\pi}{3}<\psi<\pi$. Using the asymptotic representation of the function $w_1(s)$, we readily verify that the integral (4.1) converges on the given contour and that, in so doing, for large ω, $x>a$, and $z>0$ it admits an approximate calculation.

We shall assume that the source is located sufficiently far from the boundary $z=0$ and that the point of observation lies in the deep shadow zone. Then, applying customary procedure for the computation of integrals along a cut with a large parameter in the exponential, we obtain

$$U_2^{(1)} = \frac{v(0)}{4\pi^{3/2}\omega^{1/3}}\frac{\varphi^2(0)}{\varphi(x)}\,\frac{e^{i\omega\int_0^x\frac{dx}{v(x)}+\frac{2i}{3}\left(\zeta_0^{3/2}+\zeta^{3/2}\right)+\frac{\pi i}{4}}\left[1+O(\omega^{-1/3})\right]}{\left[w_1'(0)\right]^2(\zeta_0\zeta)^{1/4}R^{3/2}} \tag{4.2}$$

Here

$$R = \frac{\omega^{1/3}}{2} \int_0^x v(x)\varphi^2(x)dx - \sqrt{\varsigma_0} - \sqrt{\varsigma} > 0. \tag{4.3}$$

If we compare $U_2^{(1)}$ with $U_2^{(0)}$ from (3.17), we can verify that in the given domain $U_2^{(0)}$ is exponentially small relative to $U_2^{(1)}$. Consequently, it is required in (2.14) to set $\alpha = 0$ and $\beta = 1$ and then to investigate only $U_2^{(1)}$.

In the formulation of the solution (2.14) in the boundary layer (2.1) the velocity $v_2(x, z)$ was approximately assumed equal to $v(x)[1 - \varepsilon(x)z]$, introducing an error of order $o(\omega^{-1/3})$ in the solution. If in (4.2) we perform the substitution

$$z = \frac{v(x) - v_2(x, z)}{\varepsilon(x)v(x)} \tag{4.4}$$

and make use of the expressions for ς_0 and ς from (2.3) and (3.3), we obtain within the same error limits

$$U_2^{(1)} = \frac{\sqrt{2v(a)}}{\pi^{3/2}} \frac{\varphi(a)}{\varphi(b)} \sqrt{\frac{\varphi'(b)}{\varphi(a)}} \left[\frac{\sqrt{v(b)}}{2\varepsilon(b)}\right] \frac{e^{i\omega\tau + \frac{\pi i}{4}}\left[1 + o(\omega^{-1/3})\right]}{\omega^{7/6}[w_1'(0)]^2 \sqrt{a(x\ b)}\, \tilde{R}^{3/2}}. \tag{4.5}$$

It turns out in this case that

$$\tau = \int_0^x \frac{dx}{v(x)} + \frac{2}{2\omega}\left(\varsigma_0^{3/2} + \varsigma^{3/2}\right) \tag{4.6}$$

corresponds to the propagation time of a disturbance from the source to the point of observation, a is the point of contact with the boundary $z = 0$ of a ray emanating from the source, and b is the point of the boundary at which the ray becomes detached from it (Fig. 2). The quantity $\tilde{R}$ has the form

$$\tilde{R} = \int_a^b \left[\frac{2\varepsilon(x)}{\sqrt{v(x)}}\right]^{2/3} dx \tag{4.7}$$

and describes the variation of the field on the interval (a, b).

Equation (4.5) can be used to describe the field not only for points satisfying inequality (2.1), but also for any z in the deep shadow zone. For this it is sufficient according to given $v_2(x, z)$ to calculate the quantities a, b, τ, and $\tilde{R}$.

Expression (4.5), which describes the principal part of the disturbance in the shadow zone, has a structure similar to the analogous expression for a head wave striking a second-order (strong) interface. The essential difference between them is in the power of ω in the denominator. For a head wave it is equal to $3/2$.

By integration over ω we can obtain from (4.5) an expression for the field in the shadow zone in the nonsteady case. For velocities v_1 and v_2 independent of the coordinate x this expression has been deduced from the exact solution of the problem in [2], thus affording us with indirect justification for Eq. (4.5).

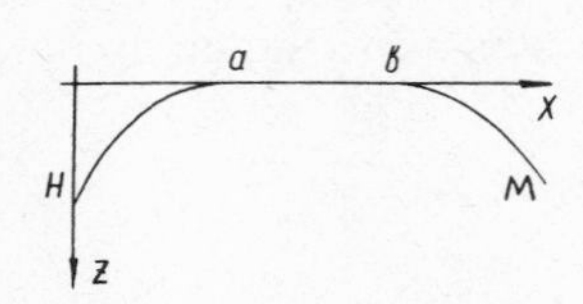

Fig. 2.

In conclusion the author expresses his appreciation to V. M. Babich for considerable aid in the investigation, as well as to I. A. Molotkov for a painstaking review of the manuscript.

LITERATURE CITED

1. Babich, V. M., A formal technique for formulating the short-wave asymptotic behavior of the Green function, Trudy Mat. Inst. Akad. Nauk SSSR, Vol. 98 (in press).
2. Tsepelev, V. N., On the propagation of oscillations in a medium with a transition layer, Trudy Mat. Inst. Akad. Nauk SSSR, 95:184-212 (1968).